Learning Country in Landscape Architecture

David S. Jones
Editor

Learning Country in Landscape Architecture

Indigenous Knowledge Systems, Respect
and Appreciation

Editor
David S. Jones
Wadawurrung Traditional Owners Aboriginal Corporation
Geelong, VIC, Australia

ISBN 978-981-15-8878-5   ISBN 978-981-15-8876-1   (eBook)
https://doi.org/10.1007/978-981-15-8876-1

Cover pattern © Melisa Hasan

This Palgrave Macmillan imprint is published by the registered company Springer Nature Singapore Pte Ltd.
The registered company address is: 152 Beach Road, #21-01/04 Gateway East, Singapore 189721, Singapore

WARNING

Aboriginal and Torres Strait Islander readers are advised that the following text may contain voices of people who have died.

FOREWORD

With European contact, an unbroken line of civilisation and culture, vested in the First Peoples of this land, was challenged and disturbed by the notion of '*Terra Nullius*'. This notion and the attendant policies it spawned over decades have created a gap of third-world proportions that despite successive years of work is still to be arrested, yet closed.

That is why '*Recasting Terra Nullius*', through its findings and recommendations, presents as a lighthouse in the shadow of an education sector that still practices, albeit in a variety of forms, 'Intellectual Nullius' that relegates Aboriginal Knowledge to the fringe.

Of all the disciplines there are none more congruent than the link between landscape architecture and the close affinities it has to the Indigenous concept of '*Country*'. This involves both cultural and curatorial obligations and responsibilities that redefine the relationship of land and place.

As educational settings across the nation struggle with Indigenous Knowledge Systems, we have in parallel witnessed over the last twenty-five years the rise of a 'black academy'. In the same stroke, in the broader academy the elevation of the probity of Indigenous Knowledge has been recognised. The ongoing question remains: what is the quality of that engagement and from where are valid examples drawn?

It is amidst this educational evolution that the SRL733 project presents itself as a major exemplar for Australian tertiary education to learn from. Through its engaging holistic perspective, by including ethical, moral, cultural and spiritual perspectives and by removing your shoes to feel 'Mother

Earth', it offers a new learning experience and a respectful sharing of values and knowledge.

For those in the tertiary education sector ever puzzled by the notion of Indigenous Knowledge Systems, this project will be a great starting point.

To this end, this is why this book is so instrumental to Australian tertiary education in offering a possible lighthouse beam as to how to proceed into the future.

Geelong, VIC, Australia   Mark Rose

ACKNOWLEDGEMENT TO COUNTRY

We would like to acknowledge the *Wadawurrung* people who are the Traditional Owners of this *Country* upon which this article and its substance originates, and that has been used as an education venue. We pay our respects to their spirit and passion in their past and present custodianship of this *Country*, including its lands, waters, skies and its terrestrial and aquatic inhabitants, and thank them for their sharing and participation in this educational journey. In addition, we would also like to pay our respects to the Elders, both past and present, of the *Kulin* Nation and extend that respect to other Aboriginals and Torres Strait Islander peoples within this readership.

> *You have been here for 200 years and look how much of my Country you white-fellas have stuffed up. We have been here for 60,000 years, and we are now attempting to fix your mistakes so that we can be here on our Country for the next 60,000 years. You guys need to learn better.* (Bell pers. comm., 2012)

ACKNOWLEDGEMENTS

Over the passage of time of the development and servicing of SRL733, the following people are gratefully acknowledged: Belinda Allwood, Veronica Arbon, Estelle Barrett, Damein Bell, Susan Bird, Wendy Brabham, *N'Arweet* Carolyn Briggs, Liz Cameron, Brenda Cherednichenko, Philip A Clarke, Paul Davis, Marcia Devlin, Shaneen Fantin, Beth Gott, Elizabeth Grant, Scott Heyes, Danièle Hromek, Liz Johnson, Siobhan Lenihan, Darryl Low Choy, *Aunty* Lyn McInnes, John Morieson (dec.), Sue Nunn, *Uncle* Lewis O'Brien, Sandy O'Sullivan, Isobel Paton, Gary Presland, Grant Revell, *Uncle* Paddy Roe (dec.), Phillip Roös, *Uncle* Kenny Saunders, Gheran Steel, Denis Rose, Mark Rose, David Rowe, Norm Sheehan, *Tandop* David Tournier (dec.), Richard Tucker, and Tyson Yunkaporta.

PROTOCOLS

This research includes generic end-of-Trimester participant unit evaluation data enveloped within Deakin University's human ethics protocols for all students who consented to at the time of their enrolment, and an additional Human Ethics approved application by the Deakin University Human Research Ethics Committee (DUHREC) entitled 'SRL733 Indigenous Narratives and Processes' coded 2017–135 dated 16 June 2017.

This research, entitled 'SRR733 Indigenous Narratives and Processes', is subject to an approved Cultural Heritage Permit WAC-P0031 issued by the Wathaurung Aboriginal Corporation in accordance with s.36(1) of the Victorian *Aboriginal Heritage Act 2006* dated 28 August 2019.

CONTENTS

1 **Introduction: Surveying the Australian Landscape** 1
David S. Jones, Kate Alder, Shivani Bhatnagar, Christine
Cooke, Jennifer Dearnaley, Marcelo Diaz, Hitomi Iida, Anjali
Madhavan Nair, Shay-lish McMahon, Mandy Nicholson, Gavin
Pocock, Uncle Bryon Powell, Gareth Powell, Sayali
G. Rahurkar, Susan Ryan, Nitika Sharma, Yang Su, Saurabh
V. Wagh, and Oshadi L. Yapa Appuhamillage

2 *Country* 11
David S. Jones, Kate Alder, Shivani Bhatnagar, Christine
Cooke, Jennifer Dearnaley, Marcelo Diaz, Hitomi Iida, Anjali
Madhavan Nair, Shay-lish McMahon, Mandy Nicholson, Gavin
Pocock, Uncle Bryon Powell, Gareth Powell, Sayali
G. Rahurkar, Susan Ryan, Nitika Sharma, Yang Su, Saurabh
V. Wagh, and Oshadi L. Yapa Appuhamillage

3 **Indigenous Knowledge Systems and Education in Australia** 19
David S. Jones, Kate Alder, Shivani Bhatnagar, Christine
Cooke, Jennifer Dearnaley, Marcelo Diaz, Hitomi Iida, Anjali
Madhavan Nair, Shay-lish McMahon, Mandy Nicholson, Gavin
Pocock, Uncle Bryon Powell, Gareth Powell, Sayali
G. Rahurkar, Susan Ryan, Nitika Sharma, Yang Su, Saurabh
V. Wagh, and Oshadi L. Yapa Appuhamillage

4 Professional Accreditation Knowledge and Policy Context 45
David S. Jones, Kate Alder, Shivani Bhatnagar, Christine
Cooke, Jennifer Dearnaley, Marcelo Diaz, Hitomi Iida, Anjali
Madhavan Nair, Shay-lish McMahon, Mandy Nicholson, Gavin
Pocock, Uncle Bryon Powell, Gareth Powell, Sayali
G. Rahurkar, Susan Ryan, Nitika Sharma, Yang Su, Saurabh
V. Wagh, and Oshadi L. Yapa Appuhamillage

5 Learning Environments and Contexts 61
David S. Jones, Kate Alder, Shivani Bhatnagar, Christine
Cooke, Jennifer Dearnaley, Marcelo Diaz, Hitomi Iida, Anjali
Madhavan Nair, Shay-lish McMahon, Mandy Nicholson, Gavin
Pocock, Uncle Bryon Powell, Gareth Powell, Sayali
G. Rahurkar, Susan Ryan, Nitika Sharma, Yang Su, Saurabh
V. Wagh, and Oshadi L. Yapa Appuhamillage

6 Student and Graduate Voices 89
David S. Jones, Kate Alder, Shivani Bhatnagar, Christine
Cooke, Jennifer Dearnaley, Marcelo Diaz, Hitomi Iida, Anjali
Madhavan Nair, Shay-lish McMahon, Mandy Nicholson, Gavin
Pocock, Uncle Bryon Powell, Gareth Powell, Sayali
G. Rahurkar, Susan Ryan, Nitika Sharma, Yang Su, Saurabh
V. Wagh, and Oshadi L. Yapa Appuhamillage

7 Respecting *Country* and People: Pathways Forward 113
David S. Jones, Kate Alder, Shivani Bhatnagar, Christine
Cooke, Jennifer Dearnaley, Marcelo Diaz, Hitomi Iida, Anjali
Madhavan Nair, Shay-lish McMahon, Mandy Nicholson, Gavin
Pocock, Uncle Bryon Powell, Gareth Powell, Sayali
G. Rahurkar, Susan Ryan, Nitika Sharma, Yang Su, Saurabh
V. Wagh, and Oshadi L. Yapa Appuhamillage

Index 117

NOTES ON CONTRIBUTORS

Kate Alder a MLArch graduate of Deakin University, is working as a strategic planner at the City of Maribyrnong in Melbourne.

Shivani Bhatnagar a MLArch graduate of Deakin University, is working as a landscape architect at the practice of Moir Landscape Architecture in Newcastle.

Christine Cooke is a MPlan (Prof) graduate of Deakin University and a doctoral candidate there, researching '*Indigenous Knowledge awareness guidance provision and fluency in Australian planning education*'.

Jennifer Dearnaley is a MLArch and PhD graduate of Deakin University. Her PhD thesis is entitled *Wadawurrung Ethnobotany as synthesised from the research of Louis Lane*; she is the co-author of the refereed article *Aboriginal uses of seaweeds in temperate Australia*, and Director of the practice Balyang Consulting in Geelong.

Marcelo Diaz is a MLArch graduate of Deakin University, and is working as a landscape architect at MINT Pool and Landscape Design in Melbourne.

Hitomi Iida a MLArch graduate of Deakin University, is working as a landscape architect at the practice of Kihara Landscapes in Melbourne.

David S. Jones oversighting Strategic Planning and Urban Design at the Wadawurrung Traditional Owners Aboriginal Corporation, is Adjunct Professor with Griffith University's Cities Research Institute and a Professor (Research) with Monash University's Monash Indigenous Studies Centre. With academic and professional qualifications in urban planning, landscape architecture and cultural heritage, he has taught, researched and published extensively across these discipline areas in the last 30 years, including in Indigenous Knowledge Systems. He has been involved with the *Victoria Square / Tarntanyangga Regeneration Project* (2017), the *Adelaide Park Lands and Squares Cultural Landscape Assessment Study* (2007), the Museum Victoria's Forest Gallery (1995–1996), the *North Gardens Indigenous Sculpture Landscape Master Plan* (2019), *Geelong's Changing Landscape* (2019), *Re-casting Terra Nullius Blindness* (2017), and has co-contributed chapters to the *Routledge Handbook to Landscape and Food* (2018), *The Handbook of Contemporary Indigenous Architecture* (2018), *Routledge Handbook on Historic Urban Landscapes of the Asia-Pacific* (2020), and *Heal the Scar – Regenerative Futures of Damaged Landscapes* (2020).

Shay-lish McMahon is a *Awakabal* woman and a MLArch graduate of Deakin University, is working as the Indigenous Services Advisor for the practice of GHD Woodhead in Melbourne, and is co-author of a refereed conference paper entitled *Aboriginal voices and inclusivity in Australian land use Country planning.*

Anjali Madhavan Nair a MLArch graduate of Deakin University, is working as a landscape architect at the practice of Ground Ink in Sydney.

Mandy Nicholson is a *Wurundjeri* woman, a PhD candidate at Monash University researching *Being on Country off Country*, and the co-author of several refereed publications including *Wurundjeri-al Biik-u (Wurundjeri Country), Mag-golee (Place), Murrup (Spirit) and Ker-up-non (People): Aboriginal living heritage in Australia's urban landscapes.*

Gavin Pocock is a sessional teacher at Deakin University, and is the author of a refereed conference paper entitled *Redeeming Fire: The use of fire as a design tool in the Australian landscape.*

Uncle Bryon Powell is a *Wadawurrung* Elder, formerly Chief Executive Officer and Chair of the Wadawurrung (Wathaurung Aboriginal Corporation) Board, [now Wadawurrung Traditional Owners Aboriginal Corporation], a former sessional teacher at Deakin University, and co-author of the refereed chapter entitled *Welcome to Wadawurrung Country.*

Gareth Powell is a *Wadawurrung* man, a ACT-based barrister, a Board member of the Wadawurrung Traditional Owners Aboriginal Corporation, and co-author of the refereed article *Kim-barne Wadawurrung Tabayl: You are in Wadawurrung Country.*

Sayali G. Rahurkar is a MLArch graduate of Deakin University, and is a landscape architect working at Inspiring Place in Hobart.

Mark Rose is a *Gunditjmara* man of western Victoria. With a thirty-five year career in education, he has contributed to a broad range of educational settings within the State, nationally and internationally. He has consulted regularly with Indigenous and non-Indigenous organisations, both nationally and internationally. For over a decade, Mark taught in predominantly postgraduate programmes at RMIT University's Faculty of Business. He has also taught in Australia, as well as in Beijing, Hong Kong, Singapore and Malaysia. At a state and national level, and with community endorsement, Mark has sat on five ministerial advisory committees. In 2003–2005 he co-chaired the *Victorian Implementation Review of Royal Commission into Aboriginal Deaths in Custody.* He assumed the position of Chair of Indigenous Knowledge Systems at Deakin University in 2008 and progressed the proposition of Indigenous Knowledge as a knowledge system. As a ministerial appointment, Mark was Chair of Batchelor Institute of Tertiary Education's Council till 2016 and now serves on the Australian Film Television and Radio School Council and Academic Board. Between 2013 and 2017, Mark was Executive Director, Indigenous Strategy and Education at La Trobe University, before returning to RMIT University as a Professor of Management, and thereupon re-joining Deakin University as Pro Vice-Chancellor, Indigenous Strategy and Innovation in 2020.

Susan Ryan is a MPlan(Prof) graduate of Deakin University, a doctoral candidate researching *Deconstructing the colonial view of Wadawurrung Country; knowledge drawn from John Wedge's Field Books of 1835–1836,* and the author of a refereed conference paper entitled *Wadawurrung Landscapes in the Victoria Planning Processes.*

Nitika Sharma is a MLArch graduate of Deakin University and is working as a landscape architect at the practice of Mexted Rimmer in Geelong.

Yang Su is a MLArch graduate of Deakin University and is working as a landscape architect at Melbourne practice.

Saurabh V. Wagh is a MLArch graduate of Deakin University, and working as a landscape architect at the practice of Moir Landscape Architecture in Newcastle.

Oshadi L. Yapa Appuhamillage is a MLArch graduate of Deakin University and is working as a landscape architect at Thomson Hay Landscape Architects in Ballarat.

Abstract

Australian tertiary education has little engaged with Indigenous peoples and their Indigenous Knowledge Systems, and the respectful translation of their Indigenous Knowledge Systems into tertiary education learning. In contrast, while there has been a dearth of discussion and research on this topic pertaining to the tertiary sector, the secondary school sector has passionately pursued this topic. There is an uneasiness by the tertiary sector to engage in this realm; it is already overwhelmed by the imperatives of the Commonwealth's 'Closing the Gap' initiative to advance Aboriginal and Torres Strait Islander tertiary education successes and appointments of Indigenous academics. As a consequence, the teaching of Indigenous Knowledge Systems relevant to professional disciplines, particularly landscape architecture where it is most apt, is overlooked and similarly scarcely addressed in the relevant professional institute education accreditation standards. This book considers this context, and a unique experimental unit hosted at Deakin University within their professionally accredited *Master of Landscape Architecture* course, offering insights as to its pedagogical rationale and structure, and an appraisal of its learning outcomes and larger impact upon the course's students and graduates, and to professionally accredited landscape architects generally in Australia and internationally.

Keywords Indigenous Knowledge Systems • Australia • Landscape Architecture Education • Built-Environment • Decolonisation • *Wadawurrung*

LIST OF FIGURES

Fig. 2.1   'Welcome to Wadawurrung Country', sign, Princes Freeway,
            Werribee South, Vic. (Image author: DS Jones, 2015)   14
Fig. 2.2   Bunjil and his brothers, cave painting, Black Range, near
            Stawell, Vic. (Image author: DS Jones, 2018)   15
Fig. 5.1   *barrangal dyara* (skin and bones) art installation, authored by
            Jonathan Jones, The Royal Botanic Gardens Sydney,
            NSW. (Image author: DS Jones 2016)   68
Fig. 5.2   Ballarat Indigenous Playground, designed by Jeavons Landscape
            Architects with Billy Blackall, Lake Wendouree, Ballarat, Vic.
            (Image author: DS Jones 2016)   69
Fig. 5.3   Barak Building, designed by ARM Architecture featuring
            William Barak's face (or the face of Traditional *ngurungaeta*
            (Elder) William Beruk of the *Wurundjeri-Wilam*), Melbourne,
            Vic. (Image author: DS Jones, 2018)   70
Fig. 5.4   Barak Building looking from the Shrine plaza, designed by
            ARM Architecture featuring William Barak's face (or the face of
            Traditional *ngurungaeta* (Elder) William Beruk of the
            Wurundjeri-Wilam), Melbourne, Vic. (Image author: DS
            Jones, 2018)   71
Fig. 5.5   Birrarung Marr, master design by Taylor Cullity Lethlean
            (TCL) in collaboration with Paul Thompson and Swaney
            Draper, featured sculpture *'Birrarung Wilam'* designed by
            Treahna Hamm, Lee Darroch and Vicki Couzens in
            collaboration with Mandy Nicholson and Glenn Romanis,
            Melbourne, Vic. (Image author: DS Jones, 2014)   72

Fig. 5.6   Forest Gallery, Museum Victoria, Taylor Cullity Lethlean
            (TCL) in collaboration with Paul Thompson and Mark Stoner,
            Melbourne, Vic. (Image author: DS Jones, 2014)   73
Fig. 5.7   Reconciliation Place, designed by Kringas Architecture in
            collaboration with Sharon Payne, Alan Vogt, Amy Leenders,
            Agi Calka and Cath Eliot, with various artists and authors (Judy
            Watson, Michael Hewes, Matilda House, Vic McGrath, Joseph
            Elu, Kevin O'Brien, Jennifer Marchant, Simon Kringas, Sharon
            Payne, Marcus Bree, Benita Tunks, Graham Scott-Bohanna,
            Andrew Smith, Karen Casey, Cate Riley, Darryl Cowrie, Archie
            Roach, Lou Bennett, Pep Gascoigne, Thancoupie Gloria
            Fletcher, Jerko Starcevic, Kev Carmody, Paul Kelly, Belinda
            Smith, Rob Tindal, Munnari John Hammond, Alice Mitchell
            Marrakorlorlo, Wenten Rubuntja, Mervyn Rubuntja, Bill
            Neidjie, Djerrkura family, Paddy Japaljarri Stewart, and Cia
            Flannery, Canberra, ACT. (Image author: DS Jones, 2019)   74
Fig. 5.8   Recreated Tyrendarra stone house, Tyrendarra Indigenous
            Protected Area, Gunditj Mirring Aboriginal Traditional Owners
            Corporation, Tyrendarra, Vic. (Image author: DS Jones, 2015) 75
Fig. 5.9   Brambuk Living Cultural Centre, *Budja Budja* / Halls Gap,
            designed by Greg Burgess Architects (Greg Burgess, David
            Mayes, Deborah Fisher, Simon Harvey, Des Cullen, Peter Ryan,
            Anthony Capsalis) Gariwerd / The Grampians, Vic. (Image
            author: DS Jones 2018)   76
Fig. 6.1   Cross-section in Restoration of Yollinko Park, designed by
            Jennifer Dearnaley, Highton, Vic. (Image author: Jennifer
            Dearnaley 2015a, b)   105
Fig. 6.2   Masterplan design for proposed Wathaurong Cultural Centre,
            Anakie, Vic. (Image author: Brian Tang 2013)   106
Fig. 6.3   *Anakie Youang* / The Anakies [comprising *Coranguilook*,
            Baccheriburt and *Woollerbeen*], Anakie, Vic. (Image author:
            Susan Ryan 2017)   107

List of Tables

Table 1.1   Findings and Recommendations of the *Re-casting Terra Nullius Blindness: Empowering Indigenous Protocols and Knowledge in Australian University Built Environment Education* (Jones et al., 2017) research project   4

Table 4.1   Current Recognised Tertiary Education Quality and Standards Agency (TEQSA) University providers in Australia and whether they host professionally-accredited built environment courses   47

Table 4.2   Key Australian architecture projects   53

Table 4.3   Key Australian landscape architecture projects   53

Table 4.4   Key Australian urban/regional planning projects   54

Table 4.5   Position statement: Connection to country—case studies   55

Table 5.1   Deakin University's Graduate Learning Outcomes (GLOs)   64

Table 5.2   Select Australian university Graduate attributes (or GLOs) focused on Aboriginal and Torres Strait Islander peoples   64

Table 5.3   SRL733s Aims and Learning Objectives   77

Table 5.4   SRL733s Unit Learning Outcomes (ULOs)   78

Table 5.5   SETU and eVALUate generic data for SRL733   80

Table 5.6   SETU and eVALUate specific data for SRL733   81

Table 6.1   Q1: What was the most important ideas and or information that you learnt in the SRL733 unit?   92

Table 6.2   Q2: Has/Did the unit SRL733 influenced your choice of
            Masterclass and or Thesis topic(s)?   96
Table 6.3   Q3: Has this unit SRL733 changed your perspective about
            Indigenous communities, their values, and their Knowledge
            Systems? And how?   99
Table 6.4   Q4: Thinking as a graduate / practitioner, how important is
            the SRL733 learning content for your work activities?   103

## CHAPTER 1

Introduction: Surveying the Australian Landscape

David S. Jones, Kate Alder, Shivani Bhatnagar,
Christine Cooke, Jennifer Dearnaley, Marcelo Diaz,
Hitomi Iida, Anjali Madhavan Nair, Shay-lish McMahon,
Mandy Nicholson, Gavin Pocock, Uncle Bryon Powell,
Gareth Powell, Sayali G. Rahurkar, Susan Ryan,
Nitika Sharma, Yang Su, Saurabh V. Wagh,
and Oshadi L. Yapa Appuhamillage

D. S. Jones (✉) • U. B. Powell
Wadawurrung Traditional Owners Aboriginal Corporation,
Geelong, VIC, Australia
e-mail: davidsjones2020@gmail.com

K. Alder
Maribyrnong City Council, Melbourne, VIC, Australia

S. Bhatnagar • S. V. Wagh
Moir Landscape Architecture, Islington, NSW, Australia

C. Cooke • S. Ryan
School of Architecture & Built Environment, Deakin University,
Geelong, VIC, Australia
e-mail: ccooke@deakin.edu.au; rsusa@deakin.edu.au

In reviewing the position of Indigenous Knowledge Systems (IKS) within the education offerings and literature of Australian tertiary built environments (architecture, landscape architecture, urban planning), several conclusions can be drawn. In contrast to major secondary school exposure

J. Dearnaley
Balyang Consulting, Geelong, VIC, Australia

M. Diaz
MINT Pool and Landscape Design, Melbourne, VIC, Australia
e-mail: marcediaz9940@gmail.com

H. Iida
Kihara Landscapes, Melbourne, VIC, Australia

A. M. Nair
Ground Ink, Sydney, NSW, Australia

S.-l. McMahon
GHD Woodhead, Melbourne, VIC, Australia
e-mail: Shay.McMahon@ghd.com

M. Nicholson
Tharangalk Art, Melbourne, VIC, Australia

G. Pocock
Garden Consultants, Geelong, VIC, Australia

G. Powell
Wadawurrung Traditional Owners Aboriginal Corporation,
Geelong, VIC, Australia

Legals Lawyers and Barristers (LLB) Pty Ltd, Canberra, ACT, Australia

S. G. Rahurkar
Inspiring Place, Hobart, TAS, Australia
e-mail: sayalirahurkar@gmail.com

N. Sharma
Mexted Rimmer Landscape Architecture, Geelong, VIC, Australia

Y. Su
Landscape Architect, Melbourne, VIC, Australia

O. L. Yapa Appuhamillage
Thomson Hay Landscape Architecture, Ballarat East, VIC, Australia

and immersive outreach activities and literature reports, the tertiary sector is considerably devoid of engagement and discourse. What exists in this realm, for these disciplines, is largely a collection of histories, 'identity politic', anthropology, 'Aboriginal Architecture', or sociology-based mandatory or elective units (or subjects) ostensibly addressing the topics of reconciliation, deconstructionism, feminism, design typological categorisations, the 'Frontier Wars', and an eclectic mix of semi-connected optional design studies hidden inside predominately architecture courses. The latter are immersive, experiential and involve traditionally site-specific installation constructions or pre-construction design appraisals in their approach. There is little about decolonialised learning outcomes, decolonised education experiences themselves, or the role and potential of decolonised designs in embracing traditional Indigenous listen-study-critique pedagogical approaches.

This is exceedingly worrying in an area where future built environment professional practitioners are being educated in the nuances of their disciplines and the nuances and protocols of their mainstream clients, and not all their prospective client catchment—pertinent to both Aboriginal and Torres Strait Islander peoples and international students generally.

There is also a concept that the foundations of professional accreditation in Australia assume that accreditation is targeted for students/graduates to learn/work/practice in Australia. This perceived focus does not adequately address international fluency and transferability for students who learn/work in Australia and practice overseas. This is evident in the deep financial reliance of Australian universities upon international students.

Jones et al. (2017) and Tucker et al. (2016) offer the most recent contributions in this discourse for the Australian built environment sector. These authors report on the results of a horizontal analysis of what is (and is not) transpiring in Australia's built environment professionally accredited courses—architecture, landscape architecture, urban planning—pointing to major deficiencies in this area. Their research, entitled *Recasting Terra Nullius Blindness*, and detailed more extensively in Jones et al. (2017, 2018a) provides a comprehensive analysis that is gathering peer recognition and acceptance. The findings and recommendations from this research are presented in Table 1.1.

Table 1.1 Findings and Recommendations of the *Re-casting Terra Nullius Blindness: Empowering Indigenous Protocols and Knowledge in Australian University Built Environment Education* (Jones et al., 2017) research project

Findings	Recommendations
Indigenous self-determination and capacity building lies at the core of sharing Indigenous Knowledge in built environment higher education. The project team concluded that: immediate steps should be taken to develop ethical and participatory processes to decolonise the curricula of the architecture, landscape architecture and planning programs at Australian universities; there is a shortage of qualified Aboriginal and or Torres Strait Islander staff to address suitable Indigenous Knowledge Systems transmission and related issues sufficiently in built environment higher education; there are opportunities for universities to develop consistent policies with regard to Indigenous Knowledge and peoples; there are opportunities for built environment schools to develop more content on Indigenous Knowledge Systems and associated protocols of engagement in response to student demand; there are opportunities for built environment schools to increase the amount and/or variability of content and discipline-specific knowledge being taught; built environment academic staff expressed a desire for guidance to boost their confidence when addressing Indigenous content, and referring to Indigenous peoples, and linking to the international context;	Accordingly, the following recommendations are made: **That Australian universities:** there is a need for universities to consider and heed these recommendations in escalating the execution of their policy aspirations top-down, but also to encourage quality and support bottom-up initiatives by built environment academic staff; appoint, resource and promote Aboriginal and or Torres Strait Islander academics in the built environment disciplines, and financially and academically enable dedicated masters by coursework and PhD scholarships in the built environment; develop ethical and participatory processes to decolonise built environment curricula; provide better entry, undergraduate and postgraduate pathways to enable and support Aboriginal and Torres Strait Islander student advancement in the built environment sector; nurture the maturation of generic and associated discipline-specific curricula that respectfully integrates and embeds Indigenous Knowledge Systems content and research methodologies as well as knowledge decolonisation strategies; seek to implement the now agreed recommendations of the *Report of the Review of Higher Education Access and Outcomes for Aboriginal and Torres Strait Islander People* (Behrendt et al. 2012), and the Aboriginal and Torres Strait Islander Higher Education Advisory Council, within an agreed period of time; have an approved Reconciliation Action Plan or Indigenous Strategy in place that articulates the university's aspirations to implement the now agreed recommendations of the *Report of the Review of Higher Education Access and Outcomes for Aboriginal and Torres Strait Islander People* (Behrendt et al. 2012) and Indigenous Knowledge Systems into the university community and culture; need to implement university upper level policy through changes in middle level management and policy implementation, cascading to the lower level built environment curriculum design and execution;

both practitioners and representatives of the Australian Institute of Architects (AIA), Australian Institute of Landscape Architects (AILA) and Planning Institute Australia (PIA), during the period of the investigation, interpreted 'Indigenous' to mean only 'Aboriginal and Torres Strait Islander' culture and did not recognise that this term may have other meanings for international students, and how they may be helped to deepen their appreciation and understanding of this realm;

there is an opportunity for the AIA, AILA and PIA to lead Australian development of an integrated contemporary education standards policy with regards to Aboriginal and Torres Strait Islander peoples knowledge systems and connections to *Country* and what the likely impact this has for each of the respective professions; and,

there is an opportunity for educators to produce discipline-specific knowledge to address the needs to produce employment-ready graduates, skilled in the protocols to engage with Aboriginal and Torres Strait Islander peoples and Indigenous Knowledge Systems.

recognise that significant education exemplars exist in the built environment sector, and that Indigenous Knowledge Systems can be successfully taught;

develop a support network for those academics and Indigenous students in the built environment education sector who are interested in advancing the above initiatives; and, better support the needs of incoming international students to enhance their understanding of 'Indigenous' cultures, including communities within their own nations, rather than only providing introductions to Australian Aboriginal and Torres Strait Islander cultures.

That the Architects Institute of Australia (AIA), the Australian Institute of Landscape Architects (AILA) and the Planning Institute of Australia (PIA):

foster policy and coursework accreditation standard renovations to ensure commonality of text and definitions in their respective education policies as well as alignment with the above recommendations and findings of this research;

ensure commonality of learning expectations so that students obtain an understanding of Aboriginal and Torres Strait Islander peoples and Indigenous culture generally pertinent to their discipline;

undertake better monitoring and reporting upon progress in the execution of these recommendations; and,

develop a cross-Australia support network for those practitioners in the built environment education sector who are interested in advancing the above initiatives.

That the Australian Government Department of Education and Training:

consider supporting an Indigenous learning and teaching forum in 2017 and/or a learning and teaching good practice report. This built environment-oriented project is one of a number of OLT-supported fellowships and projects that have been completed since the *Report of the Review of Higher Education Access and Outcomes for Aboriginal and Torres Strait Islander People* (Behrendt et al. 2012). In 2017, four years after the *Review*, there is an opportunity for opportunity for university academics, discipline professionals and government to explore progress and understand synergies.

Source: Jones et al. (2017: x–xii)

Recognising this research, the Australian Institute of Landscape Architects (Victorian Group) (AILA 2019) recently bestowed the *Recasting Terra Nullius Blindness* project an Award of Excellence citing:

> *Recasting Terra Nullius Blindness is a defining resource that seeks to scaffold information, advice, policy change, and recommendations that may redress and enable Indigenous knowledge systems incorporation and advancement in Australian University built environment education. A major aim of the report is to enable tertiary students in the built environment professions exposure to knowledge and cultural systems of Indigenous Australians for the enhancement of their skills in applying the appropriate protocols and processes for engaging with Indigenous Australians. Its importance cannot be understated. ... A diligent, crucial body of research and policy to enhance all built environment professions.*

Notably, Australian professional practice institutes rarely award one of their annual prizes to an academic-derived project.

Thus, a key finding was that students were eager to learn about Indigenous culture and Indigenous Knowledge Systems, but that the academic resources and fluency was wanting, apprehensive and lacked any exemplars. This book outlines an answer to the latter.

What is missing in this deep horizontal analysis is a complementary vertical analysis of a case study learning experience. This book offers this important insight drawing upon autoethnological, qualitative and quantitative data evaluating learning experiences, outcomes, and influences of a discrete postgraduate unit called 'SRL733 Indigenous Narratives and Processes' (SRL733).

This book reviews this innovative unit, located within the 2-year equivalent AILA professionally accredited S703 *Master of Landscape Architecture* (MLArch) course, at Deakin University (Deakin). The book considers the rationale, performance and learning outcomes of SRL733, offers voices of student participants, and considers the unit within the wider tertiary built environment professional accreditation regimes.

SRL733 focuses upon introducing post-graduate landscape architecture students to the protocols and cultural relationships of Indigenous peoples to their landscapes, and how they—as the future designers of our built environment—can better engage with these communities and peoples as practitioners. It is not about history, identity politics, sociology, anthropology or re-writing Australia's past 'history' or celebrating Aboriginal culture.

Rather, it is about appreciating, learning from, and understanding of an Indigenous view of environment (Rose 1996), *Country*, stories, place-making and design/planning (Pieris et al. 2014), and engagement, climate change context (Jones et al. 2018b) and offering and understanding the notion of design/planning by respect. Thus, SRL733 has strong unit learning outcomes (ULOs) with an emphasis upon environmental and cultural knowledge re-envisioning and understanding, to better comprehend and professionally navigate with their prospective clients.

Secondly, SRL733 recognises the actual profile of Australian tertiary students that are not solely Australian but a mix of Australian and international students (Jones et al. 2017). Accordingly, SRL733 is not focused solely upon Aboriginal and Torres Strait Islander peoples, as the expression 'Indigenous' often assumes the Australian mainstream definition of 'Indigenous' = 'Aboriginal and Torres Strait Islander peoples'; however, it enables the consideration of First Nations peoples globally. This strategy has been deliberate to bring both international and domestic student voices actively into the classroom, drawing from their home landscapes, because these are their future professional practice venues. Additionally, the class cohort is regularly a rich mixture of Australian, Chinese, Indonesian, Sri Lankan, Indian, Malaysian, Vietnamese, and South American students resulting in fascinating cross-cultural exchanges of values and perspectives that offer connection and comparison.

Much of the student cohort, both Australian and international, have not been confronted with different ways of thinking and knowing. Students have generally lived and studied within a dominant culture, often without consciousness of any social 'Other'. Through undertaking SRL733 students become aware of multiple world views and ways of experiencing time, space and the world around them. While specifically directed at an Indigenous world view, the learnings of acceptance and engagement with Indigenous people and questioning one's own held beliefs can also be applied when working with other non-culturally dominant sections of the community such as lesbian, gay, bisexual, and transgender (LGBTI) or refugee groups.

Geographically, the unit is taught on *Wadawurrung Country* at Deakin's Waterfront Campus (Powell et al. 2019; Powell and Jones 2018; Wadawurrung 2019). While centred in *Wadawurrung Country*, the unit is taught in both face-to-face and online modes, in second trimester, at the same time availing access to any enrolled student around Australia. Accordingly there is a strong *Wadawurrung* engagement in the unit's teachings. For 2020, the teaching strategy will remain solely online given the current COVID-19 context, and additional e-learning resources are being put in place to enliven the learnings real-time.

While SRL733 started as a mandatory unit in 2012 inside the MLArch course, it quickly had become an optional elective unit within the Planning Institute of Australia (PIA) professionally accredited S764 *Master of Planning (Professional)* (MPlan) and the H757 *Master of Public Health* courses at Deakin, and more recently a required elective within Deakin's professionally accredited S700 *Master of Architecture* (MArch) course.

As a note, the authorship of this book has been prepared by the teaching staff, sessional teaching staff, recent domestic and international students and graduates who have participated in the teaching and learning of SRL733. The authors include people with Australian, *Wadawurrung*, Sri Lankan, Chinese, Indian, Japanese, Columbian, *Wurundjeri* and *Awabakal* cultural backgrounds. The terms in italics in this book pertain to distinct Aboriginal *Country*'s in Australia or specific Aboriginal relevant language words.

To enter into this discussion, the reader needs to understand the concept of *Country*, issues about decolonization theory and application, the current academic environment about IKS in built environment tertiary education, before appreciating the nature of SRL733 and its role in re-charting values, respect, and personal reflectivity.

REFERENCES

Australian Institute of Landscape Architects (AILA). (2019). *AILA Awards and Competitions*. Canberra: Australian Institute of Landscape Architects. Accessed at: https://aila.awardsplatform.com/gallery/xDMAkbDz/lqznyYZq?search=fe703e215eea632b-91. Accessed 1 May 2020.

Behrendt, L., Larkin, S., Griew, R., & Kelly, P. (2012). Review of Higher Education Access and Outcomes for Aboriginal and Torres Strait Islander People Final Report AGPS, Canberra.

Jones, D. S., Low Choy, D., Revell, G., Heyes, S., Tucker, R., & Bird, S. (2017). *Re-casting Terra Nullius Blindness: Empowering Indigenous Protocols and Knowledge in Australian University Built Environment Education*. Canberra: Office for Learning and Teaching / Commonwealth Department of Education and Training. At http://www.olt.gov.au/project-re-casting-terra-nullius-blindness-empowering-indigenous-protocols-and-knowledge-australian-(2017–2018); at https://ltr.edu.au/resources/ID12_2418_Jones_Report_2016.pdf (2018). Accessed 1 May 2020.

Jones, D. S., Low Choy, D., Tucker, R., Heyes, S., Revell, G., & Bird, S. (2018a). *Indigenous Knowledge in the Built Environment: A Guide for Tertiary Educators.* Canberra: Office for Learning and Teaching / Commonwealth Department of Education and Training. https://ltr.edu.au/resources/ID12-2418_Deakin_Jones_2018_Guide.pdf. Accessed 1 May 2020.

Jones, D. S., Roös, P. B., Dearnaley, J., Threadgold, H., Nicholson, N., Wissing, R., Berghofer, D., Buggy, R., Low Choy, D., Clarke, P. A., Serrao-Neumann, S., Kitson, G., Ryan, S., Powell, B., Powell, G., & Kennedy, M. G. (2018b). ReCrafting Urban Climate Change Resilience Understandings – Learning from Australian Indigenous Cultures. In *Biophilia Smart Resilience: e-Proceedings of the 55th International Federation of Landscape Architects World Congress 2018* (pp. 402–417). Singapore: Marina Bay. http://www.ifla2018.com/eproceedings. Accessed 1 May 2020.

Pieris, A., Tootell, N., Johnson, F., McGaw, J., & Berg, R. (2014). *Indigenous Place: Contemporary Buildings, Landmarks and Places of Significance in South East Australia and beyond.* Carlton: Melbourne School of Design, University of Melbourne.

Powell, G., & Jones, D. S. (2018). *Kim-barne Wadawurrung Tabayl:* You are in *Wadawurrung Country. Kerb: Journal of Landscape Architecture, 26,* 22–25.

Powell, B., Tournier, D., Jones, D. S., & Roös, P. B. (2019). Welcome to *Wadawurrung* Country. In D. S. Jones & P. B. Roös (Eds.), *Geelong's Changing Landscape: Ecology, Development and Conservation* (pp. 44–84). Melbourne: CSIRO Publishing.

Rose, D. B. (1996). *Nourishing Terrains: Australian Aboriginal Views of landscape and Wilderness.* Canberra: Australian Heritage Commission.

Tucker, R., Low Choy, D., Heyes, S., Revell, G., & Jones, D. S. (2016). Re-Casting Terra Nullius Design-Blindness: Better Teaching of Indigenous Knowledge and Protocols in Australian Architecture Education. *International Journal of Technology and Design Education, 28*(1), 303–322.

Wadawurrung (Wathaurung Aboriginal Corporation). (2019). *Wadawurrung Country of the Victorian Volcanic Plains.* Ballarat: Wadawurrung (Wathaurung Aboriginal Corporation).

## CHAPTER 2

Country

David S. Jones, Kate Alder, Shivani Bhatnagar,
Christine Cooke, Jennifer Dearnaley, Marcelo Diaz,
Hitomi Iida, Anjali Madhavan Nair, Shay-lish McMahon,
Mandy Nicholson, Gavin Pocock, Uncle Bryon Powell,
Gareth Powell, Sayali G. Rahurkar, Susan Ryan,
Nitika Sharma, Yang Su, Saurabh V. Wagh,
and Oshadi L. Yapa Appuhamillage

As discussed earlier, built environment tertiary education in Australia seeks to ground future non-Indigenous designers in appreciating how to work on built environment projects with Indigenous clients and end users. Part of this grounding is the need to understand the assumptions and dominant

D. S. Jones (✉) • U. B. Powell
Wadawurrung Traditional Owners Aboriginal Corporation,
Geelong, VIC, Australia
e-mail: davidsjones2020@gmail.com

K. Alder
Maribyrnong City Council, Melbourne, VIC, Australia

S. Bhatnagar • S. V. Wagh
Moir Landscape Architecture, Islington, NSW, Australia

© The Author(s), under exclusive license to Springer Nature
Singapore Pte Ltd. 2021
D. S. Jones (ed.), *Learning Country in Landscape Architecture*,
https://doi.org/10.1007/978-981-15-8876-1_2

biases behind Euro-centric notions of built environment in comparison to Aboriginal concepts of *Country*, and in turn, how designers might work within and between these concepts. Landscape is in many ways *Country* but it holds different values. Additionally, across Australia there are over

C. Cooke • S. Ryan
School of Architecture & Built Environment, Deakin University,
Geelong, VIC, Australia
e-mail: ccooke@deakin.edu.au; rsusa@deakin.edu.au

J. Dearnaley
Balyang Consulting, Geelong, VIC, Australia

M. Diaz
MINT Pool and Landscape Design, Melbourne, VIC, Australia
e-mail: marcediaz9940@gmail.com;

H. Iida
Kihara Landscapes, Melbourne, VIC, Australia

A. M. Nair
Ground Ink, Sydney, NSW, Australia

S.-l. McMahon
GHD Woodhead, Melbourne, VIC, Australia
e-mail: Shay.McMahon@ghd.com

M. Nicholson
Tharangalk Art, Melbourne, VIC, Australia

G. Pocock
Garden Consultants, Geelong, VIC, Australia

G. Powell
Wadawurrung Traditional Owners Aboriginal Corporation,
Geelong, VIC, Australia

Legals Lawyers and Barristers (LLB) Pty Ltd, Canberra, ACT, Australia

S. G. Rahurkar
Inspiring Place, Hobart, TAS, Australia
e-mail: sayalirahurkar@gmail.com

N. Sharma
Mexted Rimmer Landscape Architecture, Geelong, VIC, Australia

Y. Su
Landscape Architect, Melbourne, VIC, Australia

O. L. Yapa Appuhamillage
Thomson Hay Landscape Architecture, Ballarat East, VIC, Australia

250 *Country*'s that each Indigenous community and or nation reside within, each with a different name, customs, linguistic structure, set of environmental protocols, very much analogous to medieval Europe with its multiplicity of principalities and duchies.

The traditional Western definition of landscape is time, place and cultural context dependent—it holds a clear place in time (Stilgoe 1980). European landscape painters between the 15th and 19th centuries influenced early Western interpretations of landscape as being predominantly rural or wilderness (Mitchell et al. 2009: 17), with an emphasis upon the depiction of natural attributes and aesthetic qualities. Non-Indigenous geographer Carl Sauer, in the twentieth century, argued for a more humanistic comprehension of landscape that embraced cultural perception, memory and use. He described landscape as '... an area made up of a distinct association of forms, both physical and cultural' (Sauer 1925: 26). This definition recognised that human interactions (history and culture) and their expression in places interplay with natural processes. The United Nations Educational, Scientific and Cultural Organisation (UNESCO), in assembling 'world heritage', has recognised the apparent dichotomy between nature and culture, thus formulating the term 'cultural landscapes' as being 'the combined works of nature and man' (Mitchell et al. 2009: 18). This is interesting because such a definition fits very uncomfortably within UNESCO's *Declaration on the Rights of Indigenous Peoples* (UNDRIP) (2007) (Figs. 2.1 and 2.2).

In contrast, Indigenous Australians perceive the natural landscape— *Country*—as a living cultural landscape, irrespective of whether it has now become 'colonised' or developed or changed to accommodate post-contact towns and cities. To them, their *Country* is alive and enriched with Indigenous cultural identity (past, current and future), memory, places and history. Canadian archaeologist Buggey (1999: 30), reflecting upon her Australian practice exposure to the Canadian Indigenous context, observes that:

> *An Aboriginal cultural landscape is a place valued by an Aboriginal group (or groups) because of their long and complex relationship with that land. It expresses their unity with the natural and spiritual environment. It embodies their traditional knowledge of spirits, places, land uses, and ecology. Material remains of the association may be prominent, but will often be minimal or absent.*

Fig. 2.1 'Welcome to Wadawurrung Country', sign, Princes Freeway, Werribee South, Vic. (Image author: DS Jones, 2015)

Thus the Western concept of a cultural landscape exists partially both within and outside of *Country*, but is not holistically *Country*. However, *Country* is more than simply a cultural landscape. As non-Indigenous

Fig. 2.2 Bunjil and his brothers, cave painting, Black Range, near Stawell, Vic. (Image author: DS Jones, 2018)

philosopher Rose (1996: 7) has eloquently stated, *Country* to Aboriginal peoples is a realm of infinite meaning, time and personality:

> *Rather, country is a living entity with a yesterday, today, tomorrow, with a consciousness, and a will toward life. Country is multi-dimensional- it consists of people, animals, plants, Dreamings; underground, earth, soils, minerals and waters, surface water, and air. There is sea country and land country; in some areas people talk about sky country. Country has origins and a future; it exists both in and through time.*

Thus, the *Mak Mak* people from *Country* south-west of Darwin, in the Northern Territory, observe that their *Country* affects their emotion and memory:

> *What's the first thing that enters your mind when you hear the name of your country? 'Matri lerri weti ngirrbuty tyen.' [I am happy now], sometimes*

referred to as having a 'good binyji' [feeling of being glad all over]. 'I can look every place. All the sites pass by me like a slide show. I see the my (old people) walking across the great plains, making camp, I can still remember their voices as they would cry out, laugh and chatter with one another... I vividly remember my childhood. Once again I am there with them [sic.] (Deveraux 1997: 72).

Country is thus alive. 'Country in Aboriginal English is not only a common noun but also a proper noun. People talk about Country in the same way that they would talk about a person [*sic*.]' (Rose 1996: 7). Thus, to forget *Country* from built environment discipline design processes about land and water is like forgetting about family.

Indigenous Australian ontologies involve ways of being and knowing that emerge from the intertwining relationships between *Country*, (Aboriginal) Law [lore] and *Dreaming*; the latter term in italics we recognize as a Westernised descriptor of this four-dimensional realm. In contrast, Western epistemologies reflect a position in which people own, divide and control land. In this lens, *Palawa* woman and architect Sarah Rees has written, 'Aboriginal people don't "own" land, they are "owned by it"' (Rees 2018: 181). Similarly Burarrwanga et al. has expressed in a *Yuin* woman, Danièle Hromek's (2018: 222) research that 'Since country cannot be divided, while the land may be damaged or traumatised by colonial processes, like a broken arm, it can heal through Country, Thus if one part is removed physically, the songs and stories keep it in place, in memory, and its knowledges remain intact [*sic*.]'.

Accordingly, while European settlement of the Australian continent and its waters involves a history of dispossession, predicated upon the overturned legal tenet of *terra nullius* (Australia 1992), it has little altered Indigenous peoples' perceptions of spaces, and their overarching concept of their *Country* remains unaffected.

As built environment practitioners and researchers, as narrated in SRL733, when we work with Indigenous people on projects, there should be direct but respectful intent that realizes and expresses our interest in talking about *Country*, and how it links with the particular project we are working on. Making this intent necessitates clear, specific, respectful and relationship-building communication that includes requesting permission to talk about *Country*.

REFERENCES

Buggey, S. (1999). *An Approach to Aboriginal Cultural Landscapes.* Available at: http://ip51.icomos.org/~fleblanc/in-memoriam/buggey-susan/im_buggey-susan_1999_Aboriginal_Cultural_Landscapes_HSMBC.pdf. Accessed 1 May 2020.

Deveraux, K. (1997). Looking at Country form the Heart. In D. B. Rose & A. Clarke (Eds.), *Tracking Knowledge in Northern Australian Landscapes* (pp. 68–81). Canberra: Northern Australia Research Institute.

Hromek, D. (2018). Always Is: Aboriginal Spatial Experience of Land and Country. In R. Kiddle, L. P. Stewart, & K. O'Brien (Eds.), *Our Voices-Indigeneity and Architecture.* Novato: ORO Editions.

Mitchell, N., Rössler, M., & Tricauld, P.-M. (2009). *World Heritage Cultural Landscapes.* Paris: UNESCO.

Rees, S. L. (2018). Closing the (non-Indigenous) Gap. In R. Kiddle, L. P. Stewart, & K. O'Brien (Eds.), *Our Voices- Indigeneity and Architecture.* Novato: ORO Editions.

Rose, D. B. (1996). *Nourishing Terrains: Australian Aboriginal Views of Landscape and Wilderness.* Canberra: Australian Heritage Commission.

Sauer, C. (1925). The Morphology of Landscape. *University of California Publications in Geography, 2,* 19–53.

Stilgoe, J. (1980). Landschaft and Linearity: Two Archetypes of Landscape. *Environmental Review, 4*(1), 2–17.

CHAPTER 3

Indigenous Knowledge Systems and Education in Australia

David S. Jones, Kate Alder, Shivani Bhatnagar,
Christine Cooke, Jennifer Dearnaley, Marcelo Diaz,
Hitomi Iida, Anjali Madhavan Nair, Shay-lish McMahon,
Mandy Nicholson, Gavin Pocock, Uncle Bryon Powell,
Gareth Powell, Sayali G. Rahurkar, Susan Ryan,
Nitika Sharma, Yang Su, Saurabh V. Wagh,
and Oshadi L. Yapa Appuhamillage

D. S. Jones (✉) • U. B. Powell
Wadawurrung Traditional Owners Aboriginal Corporation,
Ballarat East, VIC, Australia
e-mail: davidsjones2020@gmail.com

K. Alder
Maribyrnong City Council, Melbourne, VIC, Australia

S. Bhatnagar • S. V. Wagh
Moir Landscape Architecture, Islington, NSW, Australia

C. Cooke • S. Ryan
School of Architecture & Built Environment, Deakin University,
Geelong, VIC, Australia
e-mail: ccooke@deakin.edu.au; rsusa@deakin.edu.au

© The Author(s), under exclusive license to Springer Nature   19
Singapore Pte Ltd. 2021
D. S. Jones (ed.), *Learning Country in Landscape Architecture*,
https://doi.org/10.1007/978-981-15-8876-1_3

3.1   'Closing the Gap' and Aboriginal Education

Whenever mainstream academic conversations in Australia raise the topic of 'Aboriginal education', the discussions rotate around preconceived notions that Aboriginals lack quality education opportunities and frameworks, and that 'we' need to colonise their educational values and

J. Dearnaley
Balyang Consulting, Geelong, VIC, Australia

M. Diaz
MINT Pool & Landscape Design, Melbourne, VIC, Australia
e-mail: marcediaz9940@gmail.com

H. Iida
Kihara Landscapes, Melbourne, VIC, Australia

A. M. Nair
Ground Ink, Sydney, NSW, Australia

S.-l. McMahon
GHD Woodhead, Melbourne, VIC, Australia
e-mail: Shay.McMahon@ghd.com

M. Nicholson
Tharangalk Art, Melbourne, VIC, Australia

G. Pocock
Garden Consultants, Geelong, VIC, Australia

G. Powell
Wadawurrung Traditional Owners Aboriginal Corporation,
Ballarat East, VIC, Australia

Legals Lawyers and Barristers (LLB) Pty Ltd, Canberra, ACT, Australia

S. G. Rahurkar
Inspiring Place, Hobart, TAS, Australia
e-mail: sayalirahurkar@gmail.com

N. Sharma
Mexted Rimmer Landscape Architecture, Geelong, VIC, Australia

Y. Su
Landscape Architect, Melbourne, VIC, Australia

O. L. Yapa Appuhamillage
Thomson Hay Landscape Architecture, Ballarat East, VIC, Australia

knowledge systems. The assumption is also implicit that Indigenous Knowledge Systems (IKS) holds no valid position within Western education pedagogies as well as its learning outcome frameworks, and thus has no legitimate position in Western learning systems.

The immediate Australian response has been to point towards the 'Closing the Gap' initiative as the answer. Closing the Gap is an Australian Commonwealth government strategy adopted in 2008 that sought to reduce social and educational disadvantage amongst Aboriginal and Torres Strait Islanders (Australia 2020). Until 2018, the Commonwealth and state and territory governments, in partnership with the Department of the Prime Minister and Cabinet, sought to address defined targets and produce an annual report that reviewed progress against the strategies' seven targets. These targets addressed life expectancy, child mortality, access to early childhood education, literacy and numeracy at specified school levels, Year 12 attainment, school attendance, and employment outcomes, as it relates to Aboriginal and Torres Strait Islanders.

Between 2008–2018 some eleven *Closing the Gap Reports* were tabled to the Commonwealth Parliament reporting upon progress in addressing and achieving these targets (Australia 2020). The *Closing the Gap Report 2019* (Australia 2019) unfortunately reported that of these seven targets, only two—early childhood education and Year 12 attainment—had been achieved. The remaining five targets were considerably behind on expectations in the strategy. From 2019, the National Indigenous Australians Agency (NIAA) assumed responsibilities for addressing the Closing the Gap targets partnership with Indigenous Australians following the signing of *The Partnership Agreement on Closing the Gap, 2019–2029* by representatives of the National Coalition of Aboriginal and Torres Strait Islander Peak Organisations (NCATSIPO), each state and territory government, and the Australian Local Government Association (ALGA).

In terms of the education targets, the Closing the Gap strategy aims at improving education for Indigenous people with mixed success. It is, however, clear that Year 12 or equivalent completions for Aboriginal and Torres Strait Islanders in the age group of 20–24 have increased from 47.4% in 2006 to 65.3% in 2016 as a result of more Indigenous people undertaking tertiary or related vocational education courses. Statistically, Indigenous students in higher education award courses more than doubled numerically from 2006 (9329) to 2017 (19,237). However, most of the targets for education are behind, although it is evident that Year 9 numeracy was variously on track in all states and territories; that there was slightly increasing positive NAPLAN (The National Assessment

Program—Literacy and Numeracy) statistics; and, that there has been stable school attendance rates. Factors hindering addressing this target, as perceived by the authors of these annual reports include: remoteness whereby students in isolated or remote communities struggle to perform or attend as well as students residing in urban areas.

While this strategy can be applauded, and should be endorsed, it does raise several topics considered in this book. The first centres around tangible and intangible colonisation. Since the early 1800s the British occupancy of this landscape has included major educational colonisation strategies:

- major educational colonisation strategies, by governments, churches, and pseudo-social infrastructure agencies, to disenfranchise Indigenous language legitimacy, requiring adoption of a Western language (English) and erasure of Indigenous languages;
- assuming that Indigenous Knowledge Systems have no valid position in or contribution to Western educational pedagogies;
- assuming that the Australian continent was a 'natural' wilderness untouched by human occupation or management;
- offering no respect of Indigenous Knowledge Systems' accumulated knowledge and ways of seeing and knowing (and thus in designing, planning and managing landscapes, places, and etc. (including settlements, custodial burning, aquaculture engineering, etc.) are not acknowledged), and
- that as newly arrived colonisers to this landscape we have nothing to learn from the over 60,000 years of occupancy of this continent and its waters by Indigenous peoples who have witnessed and adapted to climate change, and bio-geography change as an example.

Under colonisation, this education discussion travels one way; from the power/knowledge possessor downwards. What is needed is an equal power and knowledge relationship, respectfully sharing knowledge. These threads are relevant to this book and its discussion.

Interestingly it is only recently that research has been undertaken to understand the factors that may contribute to the educational inequities between Indigenous Australian and non-Indigenous Australian students. Bodkin-Andrews et al. (2010: 227) have concluded that 'targeting domain-specific self-concepts to increase both Indigenous and non-Indigenous educational outcomes can provide potentially potent solutions for contributing to realising equitable educational standards in Australia'.

3.2   DECOLONISATION AND INDIGENOUS EDUCATION IN AUSTRALIA

In 1992, the *Mabo Case* found that 'native title' survived colonisation (Australia 1992). However, the Australian High Court also concluded that neither Indigenous sovereignty nor full title to land was recognized, and that the taking of the Australian continent was justified as 'an act of state'. In this sense, 'native title' rights were confined to 'traditional rights of ceremony, hunting and gathering'. The principles of the *Mabo Case* were made into law in the *Native Title Act 1993* (Cth) (Australia 1993), establishing a process for Aboriginals to claim their right to their own *Country* based upon tabling evidence of their relationship and associations and continuity of occupancy and custodianship. Colonisation survives the Mabo Case, as academic Irene Watson, a *Tanganekald, Meintangk* and *Boandik* woman, writes: 'the phenomenon of colonialism remains ongoing' (Watson 2015: 13).

While many First Nations peoples have turned to international law to decolonise their lives and have their right to self-determination recognized, the United Nations' *Declaration on the Rights of Indigenous Peoples* (UNESCO 2007) gives state power precedence. Under its framework, self-determination becomes what the state will allow. This process has been slow and fraught with unforeseen colonisation obstacles, part of the necessary decolonisation process, and has included discourses about reconciliation, treaties, constructional recognition and special electoral systems. In this discourse, Watson has called for 'resistance to the ongoing colonial project, which is ingrained within the education ... of the colonial state' (Watson 2015: 8). A major starting point to decolonisation, to the dismantling of the 'colonial project', begins with non-Indigenous Australians educating themselves, and being open to Indigenous Knowledge Systems. SRL733 is a part of this process.

While much has been written about the effects colonisation has had on the ways in which academics and professionals practice and associate with Australia and its lands and waters (Gammage 2011; Jackson et al. 2017; Pascoe 1997, 2014; Porter 2010, 2013; Porter and Barry 2016; Smith 1999, 2012), little of this scholarship has been devoted to the built environment disciplines. Sweet et al. (2014) conclude that colonisation has disrupted Indigenous peoples' connection to *Country*, to culture, to communities and families through policies that sought to control, stigmatise and intervene in people's lives but, indeed, much of the former exists as a layer within the veneer of Western civilisation (Powell et al. 2019).

The understanding of 'top-down development' colonisation consequences upon Aboriginal communities is little appreciated. It is also little quality voiced in a multi-disciplinary perspective like narrated by *Wadjuk* Traditional Owner academic Cheryl Kickett-Tucker et al. (2017) and her colleagues and Elders enveloped in their critique that thereupon offers practical Elder-informed strategies and tools for improving Aboriginal community development—and thus mainstream—education and community nourishment, as *Wiradjuri* academic Juanita Sherwood (1999) expresses it.

Decolonising the practice in designing and building and the drafting of policies and systems of governance and tertiary education pedagogy in Australia starts with examining how the effects of colonisation have influenced the structures and processes that create disadvantage for Indigenous people. 'Decolonising theory and practices seek to reverse these modes of being and practice to ensure all Australians uphold the rights of and understand the values of Indigenous Australian peoples' (Jones et al. 2018a: 24).

Rees, pointing to the 'Closing the Gap' initiative, observes that this initiative is about addressing Indigenous disadvantage in a 'Western system, by Western standards'. She reverses this concept and applies it to the architecture profession to ask 'how do we close the non-Indigenous gap?' (Rees 2018: 176). She suggests there is a need to increase education and cultural competency for practitioners in this space to effectively decolonise the methodological way architecture is taught and practiced.

An important explication in this discussion is intercultural design theory. This theory shifts the focus away from the object of production towards the method. It relies upon the concept of agency whereby designers are agents in an intercultural design process.

Dyirrbal gumbilbara bama woman and academic, Carroll Go-Sam, and Keys (2018: 361) have demonstrated how cultural agency can be explored in various architectural projects through collaboration, advocacy and co-governance. Go-Sam has observed that 'it is increasingly common [today] to find architectural writers and practitioners attentive to the legacies of colonized space. Some have gone further and advocated unsettling the colonial past with ambitions to decolonize the settler city' (Go-Sam 2020). Wong (2019), of CODA Studio, in reviewing contemporary art galleries and museums has called out that we 'must shift engagement with First Nations culture from something of the past, to something contemporary and continuous. … It's time to shift our focus towards the creation of spaces that act as a Corpus of Country: dynamic, interactive and ever changing collections of stories, knowledge and culture'.

Similarly, in the landscape architecture discipline, Revell et al. (2018: 473) describe the need for landscape architects working in bi-cultural contexts to develop cultural competency within a larger framework that encompasses 'a decolonised understanding of Indigenous ways of knowing'.

This infers that frameworks founded on the values and metrics of a single culture (Western, colonised) are fundamentally unsuitable for working on bicultural projects. *Saddle Lake Cree* First Nation academic, Dalla Costa (2018: 195–196), illustrates this by highlighting the inherent difficulty of accessing 'intangible, abstract, cultural undercurrents' in the context of mainstream project structures in Canada and USA. She claims that the short research desktop approach to client and place history and culture about a people, coupled with two or three community engagement sessions, negates intercultural appreciation and understanding, arguing that our cities 'are populated by diverse peoples, many of whom are underrepresented in architecture' (Go-Sam 2020). In the same vein, *Métis* Nation member and academic David Fortin has observed that the acclaimed Canadian exhibition for the 2018 Venice Architecture Biennale entitled 'Unceded' 'did not shy away from the destructive and overt practices of colonial social engineering and its ongoing repercussions in land, design, social and health inequalities' (Go-Sam 2020). This is equally the case for the majority of private and public projects for and with Indigenous people in Australia because they are often constrained by government funding and expenditure metrics and informed by mono-cultural political and policy agendas containing inflexible procedural, economic and program conditions and accountability metrics.

We would argue that increasing competency in intercultural design can be achieved through education and practice, but in a colonised nation such as Australia (or India, or Malaysia, or Sri Lanka, etc.), a deep understanding of the history, processes and ongoing effects of colonisation is essential in our education systems. This is necessary to be able to work with empathy, build relationships and to understand the place of reconciliation in the built environment.

Decolonising practices and processes would therefore seek to reverse these modes of being, and practice, and ensure all Australians uphold the rights of Indigenous peoples. For built environment academics and practitioners, the decolonisation process begins by analysing the specific roles that their own institutions have played in colonisation. Included is asking how planners, landscape architects, and architects, for example, have influenced the Australian identity; what is their generative under-standing of environmental knowledge and its stewardship; how do they and how are they ensuring that Indigenous people self-determine their rights to capacity building within their own lands and waters, and particularly within these academic and professional institutions that the Western Elders of our contemporary discreet Westernised siloed disciplines.

It might also be asked, how inclusive has it been to move away from:

> ... *problematising Indigenous peoples to a focus on strengths, capacity and resilience, and stress the importance of proper process, including allowing the time and opportunity to develop relationships and trust. Decolonising practices also include respect for Indigenous knowledge and stress the importance of reciprocity—that* [teaching], *research and practice should reflect community priorities and explicitly aim to provide useful service.* (Sweet et al. 2014: 626)

We believe that it is important to localize decolonizing practices in the fields of built environment education and professional practice. Specific Indigenous peoples and their communities have their own priorities in how these practices should be deconstructed, and improved practices implemented. Nonetheless, the seminal work of *Māori* academic Tuhiwai Smith (1999, 2012) suggests seven strategies for decolonization:

(1) *Deconstruction and reconstruction*—this involves the interrogation of how history has been incorrectly represented and includes the rewriting or retelling the stories of the past and envisioning the future to facilitate the processes of recovery and discovery;

(2) *Self-determination and social justice*—issues in teaching, research and professional practice need to identify how participants have been overpowered by Western hegemonies. Wider frameworks of thinking and practice that enact Indigenous self-determination and social justice are required.

(3) *Ethics*—principles, protocols, policies and guidelines need to be developed to protect Indigenous knowledge systems and ways of knowing;

(4) *Language*—Indigenous languages are integral to mediating the teaching, research, and community engagement processes, recov-

ering and revitalizing language, validating Indigenous knowledge and cultures of the historically marginalized and thus creating space for the inclusion of Indigenous research and practice paradigms;

(5) *Internationalisation of Indigenous Experiences*—Indigenous scholars and practitioners need to have their own spaces, local, national and international, to come together to plan, design, organize and work collectively for Indigenous self-determination;

(6) *History*—Opportunities should be provided to allow scholars and practitioners to study the past to recover or discover their history, culture and language, to enable a reconstruction or conservation of what was lost or exists that is useful to inform the present; and,

(7) *Critique*—There needs to be a continual critique of colonial influences on the academies and professions to allow Indigenous peoples to communicate from their own frame of reference.

Several Aboriginal scholars have now entered into this discourse. *Noonuccal* women Karen Martin and Booran Mirraboopa (2003: 203) question traditions of Western research into Aboriginal matters, and have proposed their own theoretical framework and methods for Indigenous and Indigenist research called *Ways of Knowing, Being and Doing*. These *Ways* include, firstly to introduce one's self to Indigenous people so that information is provided about one's cultural location and connection can be made on political, cultural and social grounds and relations established (Martin and Mirraboopa 2003: 203–204), a theme narrated also by Martin (2008, 2014) in later research. *Apalech Wik* scholar Tyson Yunkaporta recommends focusing on Aboriginal pedagogies and processes, rather than content, in the context of Aboriginal knowledge. He observes that 'working with Aboriginal pedagogies embeds Aboriginal perspectives even when specific cultural knowledge is absent' (Yunkaporta 2009: 148). This concept is important to intercultural design methodology because, even though specific Indigenous knowledge is often not shared by custodians for various reasons, a framework that preferences Indigenous ways of knowing and working will still enable culturally responsive project outcomes. Yunkaporta (2009, 2010, 2019) proposes '8 Ways of Learning':

1. *The Story*—the narrative, the middle ground between Aboriginal knowledge systems and western learning systems;
2. *The Map*—the learning journey never takes a straight path, but winds, zig-zags or goes around;

3. *The Silence*—the deepest knowledge is not in words, but the meaning behind the words, in the spaces between them, in gestures or looks, in meaningful silences, in the work of hands, in learning from journeys, in quiet reflection, in Dreaming;
4. *The Signs*—the signs and images we all see;
5. The Land—entities in the land like stones, animals, plants and rivers all provide knowledge;
6. *The Shape*—the shapes of logic from different cultural viewpoints;
7. *The Back–tracking*—in this model of learning you always give a model of the end product of any learning right at the start. This model can be broken down into increasingly smaller parts then put together—deconstructed and reconstructed. Each piece must be seen as part of the whole;
8. *The Home World*—knowledge is always centred around local communities in the region, spiralling out to national and international literature and practice.

Learning involves listening and reading. For Western learning, the process is a linear processional journey through secondary schooling and university. Teachers are now the experts in this procession; what your parents or grand-parents or great-grand-parents say is not important and not treated as education. Qualitative and quantitative data analyses, verification methods, evidence-based arguments, and constructing vertical design narratives are measured signs of your capacity to master a distinct discipline area like architecture, landscape architecture or urban planning. Gone is the student that is valued for his/her multidisciplinary renaissance capacity to analyse, navigate and abstract salient points of knowledge; albeit small pockets of students remain in philosophy and landscape architecture.

For Indigenous students, learning often involves yarning, yarning circuitously with the same story or complex stories, and yarning where parents, grand-parents, great-grand-parents and familial Elders have an acknowledged and respected expertise. Although, you might philosophically disagree with them on various points you still respected their wisdom. In this arena data is often place-centred, autoethnographically explored and learnt thereby casting aside qualitative and quantitative conventions (Kovach 2009; Martin 2008). This is a process where time and place are human and yarn back to you. Thus trees speak, water riffles warnings, plovers scream at you at the beach, songs and stories are multifaceted 'books', and time is a story/'history' of the then, the now and the future.

There is now an increasing body of literature by Indigenous academics or allied proponents discussing decolonisation of teaching and learning. These include many authors in Australia (Arbon 2008a, b; Hughes 1987; Hughes and More 1997; Hull-Milera and Arbon 2017; Martin 2008; Moran et al. 2018; Rigney 1997, 1999, 2001, 2006; Rose 2017; Sheehan 2011; Wilson 1999), in New Zealand (Bishop 1998; Bishop and Ted 1999; Smith 1999, 2012; Walker et al. 2006), in Canada (Battiste 2013; Kovach 2009; Porter and Barry 2016; Tobias 2009; Wilson 2008), in India (Ramanujan 1989; Sanyal 2012; Sen 1992; Tharoor 2015, 2017), and from other international perspectives (Chilisa 2011; Ellis 2000; Porsanger 2004) who are all increasingly questioning the assumptions and policies upon which colonisations of their landscapes occurred, offering Indigenist theoretical insights or methodological approaches, and many of these texts are used in SRL733.

To propose decolonisation is to challenge the socially dominant accepted way knowledge is today, and has been constructed in the last 200–300 years. Decolonisation involves re-envisioning the objectivity of knowledge; to be undaunted by questioning the societal norms; to recognise that to question the economic colonialism *status quo* is to question the dominant pillar of 'growth' that developed world cultures are based upon irrespective of 'sustainability'; and, a recognition that sharing and shared values are necessary for cultural humility and not 'social engineering' or nationalistic or university 'socialisation' policies.

Where Australian's built environment professional institutes sit within these knowledge and learning contexts is variable. Some alignments vary longitudinally and often on a daily basis, resulting in inconsistency, lack of commitment, and a failure to demonstrate visionary leadership ensuring that respective future generations of students are fully equipped to enter into this context.

By observing visionary changes in New Zealand and Canada in the arena of First Nations land management, landscape planning, architecture and design expression, you will witness new generations of culturally attuned architecture, landscape architecture and urban planning students who are more confident and skilled at dealing with their respective built environments than Australian students (Battiste 2013; Bishop 1998; Bishop and Ted 1999; Kovach 2009; Porter and Barry 2016; Smith 1999, 2012; Walker et al. 2006; Wilson 2008).

3.3   AUSTRALIAN EDUCATION AND INDIGENOUS KNOWLEDGE SYSTEMS

Significant events in Australian history, including recognition of native title by the Australian High Court in *Mabo v the State of Queensland [No. 2]* (Australia 1992), have heightened recognition of the rights, interests, needs and aspirations of Aboriginal and Torres Strait Islander people in Australia and internationally. Yet still not much has changed in built environment professional education to better engage with Indigenous Knowledge Systems (IKS) (as distinct from cultural competency education) (Liddle 2012; Rose and Jones 2012; Universities Australia 2011a, b). So while the aspirations of enabling a better understanding of IKS and cultural systems are embodied in the agendas of the Australian built environment professional institutes, little attempt has been made to realize this objective.

Policy and academic change is a slow process. The majority of Western scientists still view IKS as 'unscientific and lacking validity'—arguing for the status quo. Nader (1996) has noticed a shift in this view from the end of the twentieth century through the growth in research on traditional environmental knowledge (e.g., Andrews 1988; Bielawski 1984, 1990; Colorado 1988; Freeman and Carbyn 1988; Johnson 1998; Merculieff 1994; Waldram 1986). The World Conference on Science held in Budapest, in 1999, confirmed this paradigm shift, recognizing that to understand associations between different knowledge systems, governments should facilitate collaboration and built upon mutual benefit between Western scientists and the holders of traditional knowledge.

Indigenous Knowledge Systems and traditions are generationally tested, and time-tested adaptations to their unique environmental and cultural contexts (Jones et al. 2018b; Powell et al. 2019). Many UN institutions, such as the International Labour Organisation, the Food and Agriculture Organisation, Minorities Rights Group International and the World Health Organisation, and non-governmental human rights organizations including the International Work Group for Indigenous Affairs, are increasingly recognizing the need to understand and conserve IKS and traditions as valuable learning resources and utilities.

A limitation in the mainstream scientific acceptance and recognition of IKS, within the realms of Western scientific knowledge, has seen IKS' characterised as 'pseudo-science' and 'anti-science'. In response, the International Council for Science (ICSU) concluded that 'there are two systems of knowledge both of which are empirically testable and both of which are concerned with understanding and guiding practical activity within the same domain of phenomena' (Bala and Joseph 2007: 42). The ICSU concluded that a more symbiotic understanding was warranted, suggesting that IKS informs everyday local-level decision-making similar to what science does in modern life, including in the 'collection and storage of water; defence against disease and injury; interpretation of meteorological and climatic phenomena; construction and maintenance of shelter; orientation and navigation on land and sea; management of ecological relations of society and nature; adaptation to environmental/social change' (ICSU 2002: 3).

Inclusion of IKS in the Australian higher education sector has been subject to (normal academic) protracted discourse. Such is hampered because Australian universities were established in the shadow of the European model, with the knowledge taught within them reflecting this model and the dominant Anglo-Celtic/European/Western culture of the Australian nation. This is despite the fact that 'universities have a [now recognized] responsibility to ensure that they are culturally-competent and are able to engage with the communities they service' (Liddle 2012: 15), and that Aboriginal communities are undoubtedly part of these communities and the present status of Aboriginal knowledge in Australian university teaching is varied.

Such universities are all established as public (and charitable) entities under state or territory enacted legislation. Most of these acts of parliament do not mention Aboriginal and Torres Strait Islander, or Indigenous peoples. The exceptions are Victorian universities (including Melbourne, Monash, RMIT, Deakin and La Trobe (Victoria 2009a, b, c, 2010) wherein they are legally required 'to use its expertise and resources to involve Aboriginal and Torres Strait Islander people of Australia in its teaching, learning, research and advancement of knowledge activities and thereby contribute to—(i) realising Aboriginal and Torres Strait Islander aspirations; and (ii) the safeguarding of the ancient and rich Aboriginal and Torres Strait Islander cultural heritage' (Victoria 2009c: 5(f)). University performances on this legal obligation is just as minimal as the number of Aboriginal and Torres Strait Islander peoples enrolled in courses within such universities as well as being academic or professional employees of these universities.

Universities Australia (UA) serves as a common voice and advocate for Australian universities. It's *Guiding Principles for the Development of*

Indigenous Cultural Competency in Australian Universities (Universities Australia 2011a) recommends the development of programs pertaining to Indigenous cultural competency theory and practice, the need to address this in teaching and learning, and to embed IKS and perspectives in all university curricula to provide students with the knowledge, skills and understandings for 'Indigenous cultural competency', and the incorporation of Indigenous cultural competency as a formal Graduate Attribute or 'Graduate Learning Outcome' (GLO). But again, the 'scorecard' on this policy obligation is minimal.

In defense of this slowness, the 2015 Indigenous Content in Education Symposium recognized that in recent decades the built environment professions have experienced shifts towards 'a new receptivity to indigenous skills and knowledge' (Crosby et al. 2015: 1), and that 'design and architecture are important disciplines for working with and improving the lives of Indigenous Australians' (Crosby et al. 2015: 1). Such interest is evidence in growth in 'Indigenous Architecture' (Gardiner and Wells 2008; Malnar and Vodvarka 2013; UNESCO 2015), 'Indigenous Planning' (Jackson et al. 2017; Porter 2010, Porter and Barry 2016; Walker et al. 2013; Wensing 2007, 2011, 2012, 2016; Wensing and Small 2012), and architectural/placemaking expressions of Indigenous cultures (Pieris et al., 2014; McGaw and Pieris 2014).

Part of this shift is the recognition that 'aboriginal vernacular architecture is an expression of highly complex and diverse relationship between the physical, social and cosmological environment' (Memmott 2002: 3), and thus that architecture that is well-integrated with Indigenous attitudes to shelter-creation and living can help coalesce the various cultural and ecological dimensions of sustainable design (Olukanni et al. 2015). In addition, Aboriginal knowledge encompasses cultural awareness, cultural sensitivity and cultural competence—characteristics; it is suggested, that would benefit architectural education (Stewart 2015).

An increasing number of Australian organisations and government agencies are (or have done) considering how they might better respond to Indigenous heritage and culture in built environment development. More are implementing Reconciliation Action Plans (Reconciliation Australia 2017) and there is growing recognition of Indigenous values reflected in policy. An example of this is the recognition of Indigenous cultural heritage in *The Far North Queensland Regional Plan 2009–2031* (Queensland

2009: 66–71). There remains, however, a frequent disconnection between intent and actually achieving built environment outcomes that respond meaningfully to Indigenous values. This disconnection has been described in the context of landscape architecture (Jones 2002), in architecture (Grant et al. 2018; Kiddle et al. 2018; Pieris et al. 2014; Walliss and Grant 2000), in planning (Low Choy et al. 2011; Porter 2013), at length in the context of Aboriginal housing (Fien et al. 2007: 4–6, 2011; Davidson et al. 2010: 1–8; Moran et al. 2016: 2–6), and revisited in Tucker et al. (2016) and Jones et al. (2017, 2018b).

The case for developing better ways of working between cultures in the built environment includes arguments founded on principles of ethics, equity, inclusiveness, identity, authenticity and '*cultural sustainability*' (Memmott and Keys 2015: 1–2). Additionally, deep reflection on the Indigenous history of a place can help to identify authentic, place-based narratives which, with Indigenous permission and collaboration, may imbue genuine identity to new public spaces and provide greater relevance to all people. By progressing beyond entrenched Western design heuristics and digging deeper into 'an experience supported reservoir of understanding' we can create spaces that leave us feeling enlightened (Krog 1983: 76).

The adjective intercultural is defined as: 'taking place between cultures or derived from different cultures' (*Oxford Dictionary* 2018). The term intercultural design first appears in the context of architecture used by Canadian scholars Martin and Casault (2005), who described the challenges of designing for people of a culture other than one's own. The term intercultural design in relation to method in architecture practice is separate to participatory, inclusive and universal design and is distinguished by the acknowledgement of philosophical differences:

> *intercultural design is a process that brings consciousness of cultural perception, beliefs and bias to the fore, rather than containing the design focus to user participation and physiological needs.* (Fantin and Fourmile 2018: 438)

According to Memmott and Davidson (2007: 53), this field emerged from, and integrates behaviour/environment theory. While international discourse in the field of intercultural design encompasses many topics, conditions and insights, there are four key themes that emerge. These are condensed as follows:

- *Governance*: Ensure the project procurement and delivery framework and metrics encompass and value Indigenous knowledge and ways of working; uphold a balance of power, interests and representation; and commit to delivering Indigenous priorities.
- *Leadership*: Ensure there is Indigenous leadership and oversight on the project. Ensure you are working with the right people with the authority to speak for a particular place.
- *Relationships*: Seek permission to engage with Indigenous people on a project. Establish agreed protocols for communication, permissions and use of material; agree on methods that will ensure reciprocity of learning, including deep listening, and mutual exchange; acknowledge diversity; manage expectations.
- *Place*: Recognise the context of the cultural landscape that is particular to a place and its people, and understand the connections from it to the surrounding landscape.

Much has been written about differences in philosophy between Indigenous and non-Indigenous peoples in Australia. *Noonuccal Quandamookah* scholar, Martin (2008: 72–80) describes 'Ways of Knowing, Being and Doing' that are central to Indigenous epistemology Fantin and *Gimuy* Elder Gudju Gudju Fourmile (2018: 439, 450) observed that Indigenous values and ways of working often do not align with Western processes and metrics. Fourmile describes architecture projects as being defined by 'institutional learning' (Fantin and Fourmile 2018: 438), a reflection of the dominant Western foundation of knowledge that governs development. A lack of awareness of Indigenous world views and cultural protocols can lead to a perception that undertaking project work in Indigenous environments is complex to navigate.

Rees (2018: 180), a *Palawa* architecture graduate, descended from the *Plangermaireener* people of Tasmania, observed that amongst non-Indigenous professionals, not knowing where to start or how to communicate, along with fear of causing offence can create reluctance to engage on projects with Indigenous people. Rees (2018: 181) states:

> *In my opinion all architects and built environment practitioners should be capable of engaging in Indigenous projects in a sophisticated and meaningful way because everything that is constructed or deconstructed in Australia is on Aboriginal land.*

REFERENCES

Andrews, T. D. (1988). Selected Bibliography of Native Resource Management Systems and Native Knowledge of the Environment. Traditional Knowledge and Renewable Resource Management in Northern Regions. *Occasional Publication,* (23), 105–124.

Arbon, V. (2008a). Knowing from where? In A. Gunstone (Ed.), *History, Politics and Knowledge: Essays in Australian Studies* (pp. 145–146). North Melbourne: Australian Scholarly Press.

Arbon, V. (2008b). *Arlathirnda Ngurkarnda Ityirnda: Being-Knowing-Doing: De-Colonising Indigenous Tertiary Education.* Teneriffe: Post Pressed.

Australia. (1992). *Mabo v Queensland (No 2)* [1992] HCA 23, (1992) 175 CLR 1 (3 June 1992).

Australia. (1993). *Native Title Act 1993* (Cth).

Australia. (2019). *Closing the Gap Report 2019: The Annual Report to Parliament on Progress in Closing the Gap.* https://www.niaa.gov.au/sites/default/files/reports/closing-the-gap-2019/index.html. Accessed 1 May 2020.

Australia. (2020). *Closing the Gap.* Available at: https://closingthegap.niaa.gov.au/. Accessed 1 May 2020.

Bala, A., & Joseph, G. (2007). Indigenous Knowledge and Western Science: The Possibility of Dialogue. *Race & Class, 49*(1), 39–61.

Battiste, M. (2013). *Decolonizing Education: Nourishing the Learning Spirit.* Vancouver: Purich Publishing.

Bielawski, E. (1984). Anthropological Observations on Science in the North: The Role of the Scientist in Human Development in the Northwest Territories. *Arctic, 37,* 1–6.

Bielawski, E. (1990). *Cross-cultural Epistemology: Cultural Readaptation Through the pursuit of knowledge.* Edmonton: Department of Anthropology, University of Alberta.

Bishop, R. (1998). Freeing Ourselves from Neo-Colonial Domination in Research: A Maori Approach to Creating Knowledge. *International Journal of Qualitative Studies in Education, 11*(2), 199–219. Available at: https://www.tandfonline.com/doi/abs/10.1080/095183998236674. Accessed 1 May 2020.

Bishop, R., & Ted, G. (1999). Researching in Maori Contexts: Participatory Consciousness. *Journal of Intercultural Studies, 20*(2), 167–182.

Bodkin-Andrews, G., O'Rourke, V., & Craven, R. G. (2010). The Utility of General Self-Esteem and Domain-Specific Self-Concepts: Their Influence on Indigenous and Non-Indigenous Students' Educational Outcomes. *Australian Journal of Education, 54*(3), 227–305. Available at: https://research.acer.edu.au/aje/vol54/iss3/4. Accessed 1 May 2020.

Chilisa, B. (2011). *Decolonizing the interview method, in Indigenous Research Methodologies.* Thousand Oaks: Sage Publications.

Colorado, P. (1988). Bridging Native and Western Science. *Convergence, 21*(2), 49.

Crosby, A., Hromek, M., & Kinniburgh, J. (2015). *Making Space: Working Together for Indigenous Design and Architecture Curricula.* Unpublished Paper Presented at the Indigenous Content in Education Symposium 2015.

Dalla Costa, W. (2018). Metrics and Margins: Envisioning Frameworks in Indigenous Architecture in Canada. In E. Grant, K. Greenop, K. Refiti, & D. Glenn (Eds.), *The Handbook of Contemporary Indigenous Architecture* (pp. 193–221). Cham: Springer.

Davidson, J., Go-Sam, C., & Memmott, P. (2010). *Remote Indigenous Housing Procurement and Post-Occupancy Outcomes: A Comparative Study.* AHURI Positioning Paper No. 129, Australian Housing and Urban Research Institute Limited, Melbourne, https://www.ahuri.edu.au/research/position-papers/129. Accessed 1 June 2020.

Ellis, C. (2000). Auto-Ethnography, Personal Narrative, Reflexivity,: Researcher as Subject. In N. Denzin & Y. Lincoln (Eds.), *The Handbook of Qualitative Research* (pp. 733–768). Thousand Oaks: Sage.

Fantin, S., & Fourmile, G. G. (2018). Design in Perspective: Reflections on Intercultural Design Practice in Australia. In E. Grant, K. Greenop, A. L. Refiti, & D. J. Glenn (Eds.), *The Handbook of Contemporary Indigenous Architecture.* Singapore: Springer.

Fien, J., Charlesworth, E., Lee, G., Morris, D., Baker, D., & Grice, T. (2007). *Flexible Guidelines for the Design of Remote Indigenous Community Housing.* AHURI Positioning Paper. Available online at http://www.ahuri.edu.au/publications/projects/p30354. Accessed 10 Aug 2007.

Fien, J., Charlesworth, E., Lee, G., Morris, D., Baker, D., & Grice, T. (2011). Life on the edge: Housing experiences in three remote Australian indigenous settlements. *Habitat International 35*(2), 343–349.

Freeman, M., & Carbyn, L. (1988). *Traditional Knowledge and Renewable Resource Management in Northern Regions.* The University of Alberta Press.

Gammage, B. (2011). *The Biggest Estate on Earth: How Aborigines Made Australia.* Crows Nest: Allen & Unwin.

Gardiner, D., & Wells, K. (2008). *Australian Indigenous Architecture.* http://www.australia.gov.au/aboutaustralia/australian-story/austn-indigenous-architecture. Accessed 10 Aug 2017.

Go-Sam, C. (2020, May 26). Future Indigeneity: Shared Values in the Built Environment. *ArchitectureAU.* https://architectureau.com/articles/future-indigeneity/. Accessed 27 May 2020.

Go-Sam, C., & Keys, C. (2018). Mobilising Indigenous Agency Through Cultural Sustainability in Architecture: Are We There Yet? In E. Grant, K. Greenop, A. L. Refiti, & D. J. Glenn (Eds.), *The Handbook of Contemporary Indigenous Architecture* (pp. 347–380). Singapore: Springer Nature.

Grant, E., Greenop, K., Refiti, A. L., & Glenn, D. J. (Eds.). (2018). *The Handbook of Contemporary Indigenous Architecture.* Singapore: Springer.

Hughes, P. (1987). Aboriginal Culture and Learning Style—A Challenge for Academics in Higher Education Institutions. *The Frank Archibald Memorial Lecture Series* No. 2. Armitage: University of New England. Available at: https://www.une.edu.au/info-for/indigenous-matters/oorala/whats-on/frank-archibald-memorial-lecture-series/1987-frank-archibald-memorial-lecture#195853. Accessed 1 May 2020.

Hughes, P., & More, A. J. (1997). Aboriginal Ways of Learning and Learning Styles. *Paper Presented at the Annual Conference of the Australian Association for Research in Education, Brisbane, December* 4, 1997, pp. 1–56. Available at: https://www.aare.edu.au/data/publications/1997/hughp518.pdf. Accessed 1 May 2020.

Hull-Milera, V., & Arbon, V. (2017). Indigenous Knowledges: Growing Knowledges Through Research. In S. Hutchings & A. Morrison (Eds.), *Indigenous Knowledges: Proceedings of the Water Sustainability and Wild Fire Mitigation Symposia 2012 and 2013* (pp. 63–77). Adelaide: University of South Australia.

ICSU. (2002). *Science and Traditional Knowledge.* http://www.icsu.org/publications/reports-andreviews/science-traditional-knowledge/Science-traditional-knowledge.pdf. Accessed 1 May 2020.

Jackson, S., Porter, L., & Johnson, L. C. (2017). *Planning in Indigenous Australia: From Imperial Foundations to Postcolonial Futures.* London: Routledge.

Johnson, M. (1998). *Lore: Capturing Traditional Environmental Knowledge.* Collingdale: Diane Publishing.

Jones, D. S. (2002). Time, Seasonality and Design. *Paper Presented at Australian Institute of Landscape Architects National Conference: People + Place – The Spirit of Design*, Darwin, 22nd–24th August 2002.

Jones, D.S., Low Choy, D., Revell, G., Heyes, S., Tucker, R., & Bird, S. (2017). *Re-casting Terra Nullius Blindness: Empowering Indigenous Protocols and Knowledge in Australian University Built Environment Education.* Canberra: Office for Learning and Teaching / Commonwealth Department of Education and Training. At http://www.olt.gov.au/project-re-casting-terra-nullius-blindness-empowering-indigenous-protocols-and-knowledge-australian-(2017–2018); at https://ltr.edu.au/resources/ID12_2418_Jones_Report_2016.pdf (2018). Accessed 1 May 2020.

Jones, D. S., Low Choy, D., Tucker, R., Heyes, S., Revell, G., & Bird, S. (2018a). *Indigenous Knowledge in the Built Environment: A Guide for Tertiary Educators.* Canberra: Office for Learning and Teaching / Commonwealth Department of Education and Training; https://ltr.edu.au/resources/ID12-2418_Deakin_Jones_2018_Guide.pdf. Accessed 1 May 2020.

Jones, D. S., Roös, P. B., Dearnaley, J., Threadgold, H., Nicholson, N., Wissing, R., Berghofer, D., Buggy, R., LowChoy, D., Clarke, P. A., Serrao-Neumann, S., Kitson, G., Ryan, S., Powell, B., Powell, G., Kennedy, M. G. (2018b). ReCrafting Urban Climate Change Resilience Understandings – Learning from Australian Indigenous Cultures in *Biophilia Smart Resilience: e-Proceedings of the 55th International Federation of Landscape Architects World Congress 2018*, 18–21 July 2018, Marina Bay, Singapore, pp. 402–417. http://www.ifla2018.com/eproceedings. Accessed 1 May 2020.

Kickett-Tucker, C., Bessarab, D., Coffin, J., & Wright, M. (Eds.). (2017). *Mia Mia Aboriginal Community Development: Fostering Cultural Security.* Cambridge: Cambridge University Press.

Kiddle, R., Stewart, L. P., & O'Brien, K. (Eds.). (2018). *Our Voices: Indigeneity and Architecture.* Novato: ORO Editions.

Kovach, M. (2009). *Indigenous Methodologies: Characteristics, Conversations, and Contexts.* Toronto: University of Toronto Press.

Krog, S. R. (1983). Creative risk-taking. *Landscape Architecture, 73*(3), 70–76.

Liddle, C. (2012). Embedding Indigenous Cultural Competency. *Advocate: Newsletter of the National Tertiary Education Union, 19*(1), 15. Available at: https://issuu.com/nteu/docs/advocate_19_01. Accessed 1 May 2020.

Low Choy, D., Wadsworth, J., & Burns, D. (2011). Looking After Country: Landscape Planning Across Cultures. *Paper Presented at Australian Institute of Landscape Architects National Conference: Transform*, Brisbane, 11th–13th August 2011.

Malnar, J. M., & Vodvarka, F. (2013). *New Architecture on Indigenous Lands.* Minneapolis: University of Minnesota Press.

Martin, K. (2008). *Please Knock Before You Enter: Aboriginal Regulation of Outsiders and the Implications for Research and Researchers.* Teneriffe: PostPressed.

Martin, K. (2014). The More Things Change, the More They Stay the Same: Creativity as the Next Colonial Turn. In A. Reid, E. Hart, & M. Peters (Eds.), *A Companion to Research in Education.* Dordrecht: Springer Science+ Business Media.

Martin, T., & Casault, A. (2005). Thinking the Other: Towards Cultural Diversity in Architecture. *Journal of Architectural Education, 59*(1), 3–16.

Martin, K., & Mirraboopa, M. (2003). Ways of Knowing, Ways of Being and Ways of Doing: A Theoretical Framework and Methods for Indigenous and Indigenist Re-search. *Journal of Australian Studies, 76*, 203–214.

McGaw, J., & Pieris, A. (2014). *Assembling the Centre: Architecture for Indigenous Cultures: Australia and Beyond.* London: Routledge.

Memmott, P. (2002). An Introduction to Architecture and Building Traditions: Lessons from Ethno-Architects. *Paper Presented at the Additions to Architectural History: XIXth Annual Conference of the Society of Architectural Historians, Australia and New Zealand.*

Memmott, P., & Davidson, J. (2007). 'Architecture': Exploring the Treatise. *Architectural Theory Review, 12*(1), 78–91.

Memmott, P., & Keys, C. (2015). Redefining Architecture to Accommodate Cultural Difference: Designing for Cultural Sustainability. *Architectural Science Review, 58*(4), 278–289.

Merculieff, I. (1994). Western Society's Linear Systems and Aboriginal Cultures: The Need for Two-Way Exchanges for the Sake of Survival. In L. J. Ellanna & E. S. Burch (Eds.), *Key Issues in Hunter-Gatherer Research* (pp. 405–415). Oxford: Providence Berg.

Moran, M., Memmott, P., Nash, D., Birdsall-Jones, C., Fantin, S., Phillips, R., & Habibis, D. (2016). *Indigenous Lifeworlds, Conditionality and Housing Outcomes.* AHURI Final Report No. 260, Australian Housing and Urban Research Institute Limited, Melbourne, https://www.ahuri.edu.au/research/final-reports/260. Accessed 15 June 2020.

Moran, C., Harrington, G., & Sheehan, N. (2018). On Country Learning. *Design and Culture, 10*(1), 71–79.

Nader, L. (1996). *Naked Science: Anthropological Inquiry into Boundaries, Power, and Knowledge.* New York/London: Routledge.

Olukanni, D., Aderonmu, P., & Akinwumi, I. (2015). Pedagogic Repositioning of Curriculum in Architecture and Civil Engineering Education to Meet Indigenous Needs. *Paper Presented at the EDULEARN15Conference,* Barcelona.

Pascoe, B. (1997). *Wathaurong: Too Bloody Strong: Stories and Life Journeys of People from Wathaurong.* Geelong: Pascoe Publishing.

Pascoe, B. (2014). *Dark Emu, Black Seeds: Agriculture or Accident?* Broome: Magabala Books.

Pieris, A., Tootell, N., Johnson, F., McGaw, J., & Berg, R. (2014). *Indigenous Place: Contemporary Buildings, Landmarks and Places of Significance in South East Australia and Beyond.* Carlton: Melbourne School of Design, University of Melbourne.

Porsanger, J. (2004). An Essay About Indigenous Methodology. *Nordlit.* Available at: https://septentrio.uit.no/index.php/nordlit/article/view/1910/0. Accessed 1 June 2019.

Porter, L. (2010). *Unlearning the Colonial Cultures of Planning.* London: Routledge.

Porter, L. (2013). Coexistence in Cities: The Challenge of Indigenous Urban Planning in the 21st Century. In T. Jojola, D. Natcher, R. Walker, & T. Kingi (Eds.), *Walking Backwards into the Future: Indigenous Approaches to Community and Land Use Planning in the Twenty-first Century.* Montréal: McGill-Queens University Press.

Porter, L., & Barry, J. (2016). *Planning for Coexistence? Recognizing Indigenous Rights Through Land-Use Planning in Canada and Australia*. London: Routledge.

Powell, B., Tournier, D., Jones, D. S., & Roös, P. B. (2019). Welcome to *Wadawurrung* Country. In D. S. Jones & P. B. Roös (Eds.), *Geelong's Changing Landscape: Ecology, Development and Conservation* (pp. 44–84). Melbourne: CSIRO Publishing.

Queensland. (2009). *The Far North Queensland Regional Plan 2009–2031*. Brisbane: Department of Infrastructure and Planning. Available at: http://www.dlgrma.qld.gov.au/resources/plan/far-north-queensland/fnq-regional-plan-2009-31.pdf. Accessed 1 June 2019.

Ramanujan, A. K. (1989). Is There an Indian Way of Thinking? An Informal Essay. *Contributions to Indian Sociologist, 23*(41–58).

Reconciliation Australia. (2017). *Reconciliation Australia: Stretch Rap July 2017–July 2020*. Available at: https://www.reconciliation.org.au/wp-content/uploads/2019/02/reconciliation-australia-stretch-rap-2017-2020.pdf. Accessed at 1 June 2019.

Rees, S. L. (2018). Closing the (non-Indigenous) Gap. In R. Kiddle, L. P. Stewart, & K. O'Brien (Eds.), *Our Voices- Indigeneity and Architecture*. Novato: ORO Editions.

Revell, G., Heyes, S., Jones, D. S., Choy, D. L., Tucker, R., & Bird, S. (2018). Enough is Enough: Indigenous Knowledge Systems, Living Heritage and the (Re)Shaping of Built Environment Design Education in Australia. In E. Grant, K. Greenop, K. Refiti, & D. Glenn (Eds.), *The Handbook of Contemporary Indigenous Architecture* (pp. 465–493). Springer. https://doi.org/10.1007/978-981-10-6904-8_18.

Rigney, L.-I. (1997). Internationalization of an Indigenous Anticolonial Cultural Critique of Research Methodologies: A Guide to Indigenist Research Methodology and Its Principles. *WICAZO SA Review: Journal of Native American Studies, 14*(2), 109–121. Available at: https://www.jstor.org/stable/1409555?seq=1#metadata_info_tab_contents. Accessed 1 May 2020.

Rigney, L. I. (1999, Fall). Internationalisation of An Indigenous Anti-Colonial CulturalCritique of Research Methodologies. A Guide to Indigenist Research Methodology and Its Principles, in *HERDSA Annual International Conference Proceedings; Research and development in Higher Education: Advancing International Perspectives, 20*, (1997), pp. 629–636: Reprinted with permission in 1999. *WICAZO SA Review: Journal of Native American Studies, 14*(2), 109–122.

Rigney, L.-I. (2001). A First Perspective of Indigenous Australian Participation in Science: Framing Indigenous Research Towards Indigenous Australia Intellectual Sovereignty. *Kaurna Higher Education Journal*, 1–13. Available at:

https://ncis.anu.edu.au/_lib/doc/LI_Rigney_First_perspective.pdf. Accessed at 1 June 2019.

Rigney, L.-I. (2006). Indigenist research and Aboriginal Australia. In J. E. Kunnie & N. I. Goduka (Eds.), *Indigenous Peoples' Wisdom and Power: Affirming our Knowledge Through Narratives* (pp. 32–45). Aldershot: Ashgate.

Rose, M. (2017). The Black Academy: A Renaissance Seen Through a Paradigmatic Prism. In I. Ling & P. Ling (Eds.), *Methods and Paradigms in Education Research* (pp. 326–343). Philadelphia: IGI Global.

Rose, M., & Jones, D. S. (2012). Contemporary Planning Education and Indigenous Cultural Competency Agendas: Erasing Terra Nullius, Respect and Responsibility. In *Proceedings of the Australian & New Zealand Association of Planning Schools Conference* (pp. 179–187). Bendigo: La Trobe University, Community Planning and Development Program.

Sanyal, S. (2012). *Land of the Seven Rivers: A Brief History of India's Geography.* Gurgaon: Penguin.

Sen, G. (ed.) (1992). *Indigenous Vision: Peoples of India attitudes to the environment.* New Delhi: Sage Publications.

Sheehan, N. W. (2011). Indigenous Knowledge and Respectful Design: An Evidence-Based Approach. *Design Issues, 27*(4), 68–80.

Sherwood, J (1999). Community: What Is It? *Indigenous Law Bulletin, 21*; (1999) 4(19). Available at: https://www.austlii.edu.au/cgi-bin/viewdoc/au/journals/IndigLawB/1999/21.html?context=1;query=sherwood;mask_path=au/journals/IndigLawB. Accessed 15 June 2020.

Smith, L. T. (1999). *Decolonising Methodologies: Research and Indigenous Peoples* (1st ed.). London: Zed Books.

Smith, L. T. (2012). *Decolonising Methodologies: Research and Indigenous Peoples* (2nd ed.). London: Zed Books.

Stewart, P. (2015). *Indigenous Architecture Through Indigenous Knowledge: Dim sagalts' apkw nisiṁ [Together We Will Build a Village].* Vancouver: The University of British Columbia. https://open.library.ubc.ca/cIRcle/collections/ubctheses/24/items/1.0167274. Accessed 15 June 2020.

Sweet, M. A., Dudgeon, P., McCallum, K., & Ricketson, M. D. (2014). Decolonising Practices: Can Journalism Learn from Health Care to Improve Indigenous Health Outcomes? *The Medical Journal of Australia, 200*(11), 626–627. Available at: https://www.mja.com.au/journal/2014/200/11/decolonising-practices-can-journalism-learn-health-care-improve-indigenous. Accessed 1 May 2020.

Tharoor, S. (2015, India July 24). Read: Shashi Tharoor's Full Speech Asking UK to Pay India for 200 Years of Its Colonial Rule. *News 18*. Available at: https://www.news18.com/news/india/read-shashi-tharoors-full-speech-asking-uk-to-pay-india-for-200-years-of-its-colonial-rule-1024821.html.   Accessed 1 May 2020.

Tharoor, S. (2017). *Inglorious Empire: What the British Did to India.* Melbourne: Scribe Publishing.

Tobias, T. (2009). *Living Proof: The Essential Data Collection Guide for Indigenous Use and Occupancy Map Surveys.* Vancouver BC: EcoTrust Canada and Union of British Columbia Indian Chiefs.

Tucker, R., Low Choy, D., Heyes, S., Revell, G., & Jones, D. S. (2016). Re-Casting Terra Nullius Design-Blindness: Better Teaching of Indigenous Knowledge and Protocols in Australian Architecture Education. *International Journal of Technology and Design Education, 28*(1), 303–322.

UNESCO. (2007). *Declaration on the Rights of Indigenous Peoples.* Available at: https://www.un.org/development/desa/indigenouspeoples/declaration-on-the-rights-of-indigenous-peoples.html. Accessed 1 May 2020.

UNESCO. (2015). *Workshop on Revitalization of Indigenous Architectural and Traditional Building Skills.* Retrieved from Paris: http://unesdoc.unesco.org/images/0023/002337/233712e.pdf

Universities Australia (UA). (2011a). *Guiding Principles for Developing Indigenous Cultural Competency in Australian Universities.* Canberra: Universities Australia.

Universities Australia (UA). (2011b). *National Best Practice Framework for Indigenous Cultural Competency in Australian Universities.* Canberra: Universities Australia.

Victoria. (2009a). *The University of Melbourne Act 2009* (Vic).

Victoria. (2009b). *Monash University Act 2009* (Vic).

Victoria. (2009c). *Deakin University Act 2009* (Vic).

Victoria. (2010). *Royal Melbourne Institute of Technology Act 2010* (Vic).

Waldram, J. (1986). Traditional Knowledge Systems: The Recognition of Indigenous History and Science. *Saskatchewan Indian Federated College Journal, 2*(2), 115–124.

Walker, S., Eketone, A., & Gibbs, A. (2006). An exploration of Kaupapa Maori Research, Its Principles, Process and Applications. *International Journal of Social Science Research Methodology, 9*(4), 331–344. Available at: https://www.tandfonline.com/doi/abs/10.1080/13645570600916049. Accessed 1 May 2020.

Walker, R., Natcher, D., & Jojola, T. (2013). *Reclaiming Indigenous Planning.* Montreal: McGill-Queen's Press-MQUP.

Walliss, J., & Grant, E. (2000). Indigenous Education for Future Design Professionals. *Architectural Theory Review, 2*(11), 65–68.

Watson, I. (2015). *Aboriginal Peoples, Colonialism and International Law: Raw Law.* London: Routledge.

Wensing, E. (2007). Aboriginal and Torres Strait Islander Australians. In S. Thompson (Ed.), *Planning Australia: An Overview of Urban and Regional Planning.* Melbourne: Cambridge University Press.

Wensing, E. (2011). Improving Planners' Understanding of Aboriginal and Torres Strait Islander Australians and Reforming Planning Education in Australia. *Paper presented to the Third World Planning Schools Congress*, Perth, 4–8 July 2011.

Wensing, E. (2012). Aboriginal and Torres Strait Islander Australians. In S. Thompson & P. Maginn (Eds.), *Planning Australia: An Overview of Urban and Regional Planning* (2nd ed.). Melbourne: Cambridge University Press.

Wensing, E. (2016). 'Going deeper than the rhetoric': Comments on the PIA Draft Accreditation Policy for the Recognition of Australian Planning Qualifications, unpublished open letter; permission of the author granted to use/cite.

Wensing, E., & Small, G. (2012). A Just Accommodation of Customary Land Rights. *Paper presented to the 10th International Urban Planning and Environment Association Symposium*, University of Sydney, 24–27 July 2012.

Wilson, S. (1999). *Recognising the Importance of Spirituality in Indigenous Learning*. Available at: https://acea.org.au/wp-content/uploads/2015/04/Shaun-Wilson-paper.pdf. Accessed 1 May 2020.

Wilson, S. (2008). *Research Is Ceremony: Indigenous Research Methods*. Black Point, Nova Scotia: Fernwood Publishing.

Wong, K. (2019, November 18). Re-Imagining a Museum of Our First Nations. *The Conversation*. https://theconversation.com/re-imagining-a-museum-of-our-first-nations-123365. Accessed 1 May 2020.

Yunkaporta, T. (2009). *Aboriginal Pedagogies at the Cultural Interface*. Unpublished PhD Thesis, James Cook University, Townsville.

Yunkaporta, T. K. (2010). Our Ways of Learning in Aboriginal Languages. In J. Hobson, K. Lowe, S. Poetsch, & M. Walsh (Eds.), *Re-Awakening Languages: Theory and Practice in the Revitalisation of Australian's Indigenous Languages* (pp. 37–49). Sydney: Sydney University Press.

Yunkaporta, T. (2019). *Sand Talk: How Indigenous Thinking Can Save the World*. Melbourne: The Text Publishing Company.

CHAPTER 4

Professional Accreditation Knowledge and Policy Context

David S. Jones, Kate Alder, Shivani Bhatnagar,
Christine Cooke, Jennifer Dearnaley, Marcelo Diaz,
Hitomi Iida, Anjali Madhavan Nair, Shay-lish McMahon,
Mandy Nicholson, Gavin Pocock, Uncle Bryon Powell,
Gareth Powell, Sayali G. Rahurkar, Susan Ryan,
Nitika Sharma, Yang Su, Saurabh V. Wagh,
and Oshadi L. Yapa Appuhamillage

D. S. Jones (✉) • U. B. Powell
Wadawurrung Traditional Owners Aboriginal Corporation,
Geelong, VIC, Australia
e-mail: davidsjones2020@gmail.com

K. Alder
Maribyrnong City Council, Melbourne, VIC, Australia

S. Bhatnagar • S. V. Wagh
Moir Landscape Architecture, Islington, NSW, Australia

C. Cooke • S. Ryan
School of Architecture & Built Environment, Deakin University,
Geelong, VIC, Australia
e-mail: ccooke@deakin.edu.au; rsusa@deakin.edu.au

© The Author(s), under exclusive license to Springer Nature   45
Singapore Pte Ltd. 2021
D. S. Jones (ed.), *Learning Country in Landscape Architecture*,
https://doi.org/10.1007/978-981-15-8876-1_4

A summary of how many professionally accredited built environment-related tertiary courses in Australia exist today and which universities host these courses is set out in Table 4.1. The majority of courses reside in the old 'sandstone' universities across Australia rather than in the new generation of universities.

J. Dearnaley
Balyang Consulting, Geelong, VIC, Australia

M. Diaz
MINT Pool and Landscape Design, Melbourne, VIC, Australia
e-mail: marcediaz9940@gmail.com

H. Iida
Kihara Landscapes, Melbourne, VIC, Australia

A. M. Nair
Ground Ink, Sydney, NSW, Australia

S.-l. McMahon
GHD Woodhead, Melbourne, VIC, Australia
e-mail: Shay.McMahon@ghd.com

M. Nicholson
Tharangalk Art, Melbourne, VIC, Australia

G. Pocock
Garden Consultants, Geelong, VIC, Australia

G. Powell
Wadawurrung Traditional Owners Aboriginal Corporation,
Geelong, VIC, Australia

Legals Lawyers and Barristers (LLB) Pty Ltd, Canberra, ACT, Australia

S. G. Rahurkar
Inspiring Place, Hobart, TAS, Australia
e-mail: sayalirahurkar@gmail.com

N. Sharma
Mexted Rimmer Landscape Architecture, Geelong, VIC, Australia

Y. Su
Landscape Architect, Melbourne, VIC, Australia

O. L. Yapa Appuhamillage
Thomson Hay Landscape Architecture, Ballarat East, VIC, Australia

Table 4.1 Current Recognised Tertiary Education Quality and Standards Agency (TEQSA) University providers in Australia and whether they host professionally-accredited built environment courses

University nomenclature	State of Operations	Hosts an AIA-accredited Architecture degree A	Hosts an AILA-accredited Landscape Architecture degree LA	Hosts a PIA-accredited Planning degree P
Australian Catholic University Ltd	NSW Qld SA Vic			
Australian National University	ACT			
Bond University Ltd	Qld	UG + PG		UG PG
Carnegie Mellon University	SA			
Central Queensland University	Qld			
Charles Darwin University	NT			
Charles Sturt University	NSW			
Curtin University of Technology	WA	UG + PG		UG PG
Deakin University	Vic	UG + PG	PG	
Edith Cowan University	WA			
Federation University Australia	Vic			
Griffith University	Qld	UG + PG		UG PG
James Cook University	Qld			UG PG
La Trobe University	Vic			UG PG
Macquarie University	NSW			UG PG
Melbourne College of Divinity	Vic			
Monash College Pty Ltd	Vic			
Monash University	Vic	UG + PG		
Murdoch University	WA			
Navitas Bundoora Pty Ltd	Vic			
Newcastle International College Pty Ltd	NSW			
Queensland Institute of Business and Technology Pty Ltd	Qld			
Queensland University of Technology	Qld	UG + PG	UG	UG
Royal Melbourne Institute of Technology	Vic	UG + PG	PG	UG PG
South Australian Institute of Business and Technology Pty Ltd	SA			
Southern Cross University	NSW			
Swinburne University of Technology	Vic			
Sydney Institute of Business and Technology Pty Ltd	NSW			
The Flinders University of South Australia	SA			
The University of Adelaide	SA	UG + PG	PG	PG
The University of Melbourne	Vic	UG + PG	PG	PG
The University of Notre Dame Australia	WA NSW			
The University of Queensland	Qld	UG + PG		UG PG
The University of Sydney	NSW	UG + PG		PG
The University of Western Australia	WA	UG + PG	PG	UG PG
Torrens University Australia Ltd	SA			
University College London	SA			
University of Canberra	ACT	UG + PG		
University of New England	NSW			UG PG
University of New South Wales (UNSW Sydney)	NSW	UG + PG	UG	UG PG
University of Newcastle	NSW	UG + PG		
University of South Australia	SA	UG + PG		UG PG
University of Southern Queensland	Qld			
University of Tasmania	Tas	UG + PG		PG
University of Technology, Sydney	NSW	UG + PG	UG	PG
University of the Sunshine Coast	Qld			UG PG
Western Sydney University	NSW			UG PG
University of Wollongong	NSW			
Victoria University	Vic			

Each of the three professional institutes—the Architects Institute of Australia (AIA), the Australian Institute of Landscape Architects (AILA), and the Planning Institute of Australia (PIA)—entities have shifted backwards and forwards consistently over the last 25 years—often wandering directionless—in incorporating Indigenous issues and knowledge into accredited courses. Of these, architecture courses are accredited via the Architects Accreditation Council of Australia's (AACA) criterion with the AIA and relevant Registration Boards as co-partners to the process, and PIA and AILA entertain internal accreditation processes. The latest version of the *National Standard of Competency for Architects* (AACA 2020) now lacks any expectation architecture fluency or competency with Indigenous peoples and communities.

The architectural profession in Australia is additionally overseen by specific acts of parliament in Australia that establish registration boards: the NSW Architects Registration Board under the *Architects Act 2003* (NSW) (NSW 2003); Architects Registration Board of Victoria under the *Architects Act 1991* (Vic) (Victoria 1991); Board of Architects of Queensland under the *Architects Act 2002* (Qld) (Queensland 2002); The Architectural Practice Board of South Australia under the *Architectural Practice Act 2009* (SA) (South Australia 2009); Architects Board of Western Australia under the *Architects Act 2004* (WA) (Western Australia 2004); Australian Capital Territory Architects Board under the *Architects Act 2004* (ACT) (Australian Capital Territory 2004); Board of Architects of Tasmania under the *Architects Act 1929* (Tas) (Tasmania 1929); Northern Territory Architects Board under the *Architects Act 1963* (NT) (Northern Territory 1963).

In academic and practitioner architectural discourses, debates are largely about representation, symbolism, or housing. Concerns about public Indigenous architecture have been prevalent in Australia since the late twentieth century (Fantin 2003; Lochert 1997; Mallie and Ostwald 2009; Memmott 1997, 2007; Memmott and Reser 2000). Architects have employed symbolism—often abstracted references from Indigenous culture—to attach a greater level of significance to building.

Realising the outcomes of integrating these discourses takes considerable patience and comprehension of IKS that are both community and '*Country*'-specific and successfully respond to the distinct Indigenous culture of a *Country* under the guise of 'regionalism'. Sawyer (2011: 1, 26–27) has concluded that, while 'architecture has the ability to create a dialogue that will lead to improvements in understanding the culture, and thus a more harmonious relationship between Indigenous and non-Indigenous Australians,' it cannot be achieved in cultural competency curricular strategies alone. It needs to be more robustly scaffolded through built environment-specific immersive and engagement learnings and consultations. It is undeniable that the state of Indigenous housing in Australia is deplorable in comparison to non-Indigenous Australian's housing conditions (Go-Sam 2008; Nganampa Health Council Inc., et al. 1987; Pholeros 2003; Williams and Houston 1997). This is often the result of the ongoing failure of critical house hardware that severely impacts everyday living practices. But it also demonstrates a clear lack of knowledge and comprehension of contextual challenges.

The plethora of confusing national, state and local policies, government funding arrangements, medical research findings and bureaucratic machinations are also hindering culturally relevant and appropriate responses that recognise multiple issues rather than that one generic-answer-fits-all situations (Go-Sam 2008; Memmott and Chambers 2003; Pholeros 2003; Scally 2003; Tonkinson 2007; Ward 2011). Stallard (2011: 2) has concluded that 'there is no clear way to approach an Indigenous housing project' because the 'cross-disciplinary needs of Aboriginal housing leaves the architect in doubt of where to begin'. Offering a generic 'answer' in the education context is not the solution. Cultural competency curricular strategies may often be appropriate. They can also be premised upon immersive and engagement learning activities where a cursory base level of cultural awareness and competency is an assumed prerequisite but its *Country* specificity and *Country* 'teacher' teaching is absent.

Both architecture and Indigenous discourses cannot be appreciated in generic 'cultural competency' curricula, nor can they be realised in offering an 'Indigenous perspective'. They are far more complex in design theory and practice. In recognition of this, there needs to be a defined or discipline-consistent knowledge learning outcome that respective professional accreditation institutes expect a graduate to possess.

The 2019 National Architecture Conference grounded these questions in a radical reconfiguring of Australian architecture through its relationship to Indigenous knowledge. While other issues emerged during the two days, this will be the remarkable legacy of the conference. Across diverse scales of practice, we heard how engagement with the First Nations people of Australia is not only a necessity but a way to enrich and propel the design of our built environment.

A change in values was, however, evident at the 2019 National Architecture Conference where a dedicated Conference debate debated a 'radical reconfiguring of Australian architecture through its relationship to Indigenous knowledge' and more importantly that 'engagement with the

First Nations people of Australia is not only a necessity but a way to enrich and propel the design of our built environment' (Spurr 2019).

In summary, AIA progressed with their AIA Victoria Reconciliation Action Plan in 2010, now out of date, including a statement therein:

> *We imagine a country in which Aboriginal and Torres Strait Islander Australians participate in opportunities and expectancies afforded to all Australians. We imagine the maintenance, respect, celebration and growth of some of the world's oldest, richest, and most distinctive Indigenous cultures. We imagine a process where the practical and aspirational skills of architects become part of a process towards reconciliation, parity and healing. We imagine an inclusive process of reconciliation for all Australians with positive benefits for everyone. We imagine a day when Aboriginal and Torres Strait Islander cultures, histories and connectedness to Country are reflected in the built environment of our cities, regions and towns.* (AIA Vic 2010a, b: 1)

Such initiative enabled the successfully parallel evolution of the Indigenous Architecture and Design Victoria (IADV). However, the AIA has substantially been in a 'holding pattern' since this innovation in policy making and in education expectations. This is despite the Architects Accreditation Council of Australia's (AACA) *National Standard of Competency for Architects* (2020) that envelopes Indigenous Knowledge Systems within one of six 'Knowledge Domains'; the 'Social & Ethical Domain: Knowledge of the social, ethical and cultural values relevant to architectural practice and the impacts on project users and broader communities'.

Similar concerns have been expressed in relation to landscape architecture curricula by Jones (2002), Lawson and Erickson (2002) and Low Choy et al. (2011a). Sinatra and Murphy (1999) charted a now-lapsed OutReach initiative that exposed landscape architecture students to various Australian Indigenous communities and their landscape planning, management, shelter and health challenges. Revell, Heyes and Jones have continued this agenda in central WA and in south-eastern Australia respectively. Revell has undertaken annual Indigenous design studios at The University of Western Australia between a built environment Faculty and its School of Indigenous Studies since 1995. This work is well published (Revell 2002a, b, 2014; Heyes and Tuiteci 2013; Jones 2002) and has subsequently produced professional award-winning graduates. Jones (2002) has pointed to an urgent need to reappraise and incorporate Indigenous (environmental) Knowledge Systems in mainstream landscape architecture education curricula.

In this context, AILA has been moving forward over the last 25 years, taking a built environment leadership role in articulating a 'Connection to Country Policy' (AILA 2019a, b), amending its Constitution to include an 'Acknowledgement of Country' (AILA 2019c) that states:

> *We acknowledge and respect Aboriginal and Torres Strait Islander Peoples of Australia, as the traditional custodians of our lands, waters and seas. We recognise their ability to care for Country and their deep spiritual connection with Country. We honour Elders past and present whose knowledge and wisdom ensure the continuation of Aboriginal and Torres Strait Islander cultures.* (AILA 2019c: 1)

At a policy level, it has placed a tacit policy expectation that IKS is accommodated inside their accredited landscape architecture courses (AILA 2016).

In terms of the urban planning profession in Australia, while there is considerable desire to engage in this discourse, it has not generally been translated into tertiary-level execution. Wensing and Small (Wensing 2011; Wensing and Small 2012) have expressed this as a major deficiency in the tuition and grounding of future planners. Wensing's thoughts reiterate conclusions and investigations by Gurran and Phibbs (2003, 2004), who concluded that IKS and land management concepts were markedly lacking in planning education in Australia. Low Choy et al. (2010, 2011; Low Choy and Van der Valk 2011) have reinforced both conclusions but have also demonstrated the unique and valuable insights that IKS and their stakeholders can offer to conventional planning practice. Cooke's (2018) recent research in Victoria has reinforced that nothing has improved since Sarkissian's (1996) and Oberklaid's (2008) investigations.

The most recent Australian planning texts demonstrate this lack of engagement. While Thompson and Maginn (2012) and Byrne et al. (2014) both include substantive chapters on Indigenous Knowledge Systems, the more recent Brunner and Glasson (2015) contains only three mentions of the *Native Title Act 1993* (Cwth) (Australia 1993), and several disparaging mentions of Aboriginal and Torres Strait Islander peoples in the text.

In this context, PIA has similarly jumped back and forwards. It is now in a position whereby its education accreditation policy has mandatory Indigenous knowledge learning requirements (PIA 2016; Porter 2016;

Wensing 2016), but the rest of PIA's policies fail to engage in this topic despite a recent media release to the contrary that states that 'PIA reiterated its call for a First Nations Voice to be enshrined in the Constitution … but it is now time to re-double our efforts' (PIA 2019). The former embodies the following statements that expects that accredited 'Programs should be able to demonstrate that students have acquired a range of abilities that cover the following: … Operate in a manner that recognises the position of Aboriginal and Torres Strait Islander peoples as the first peoples of Australia, the relationship of those peoples to planning practice in historical and contemporary terms; and compliance to the maximum extent possible with accepted international standards of best practice working with Indigenous communities'. In 2020, PIA repeated the PIA 2019 statement with a 'call for the First Nations Voice to be enshrined in the Constitution (PIA 2020). We also support a process of agreement-making to take place between governments and First Nations. Both would serve to give Aboriginal and Torres Strait Islander peoples a genuine voice in matters that affect them, including those relating to land use planning', but made no mention of education policy change.

These disjunctures of policies, accreditation agendas and practice aspirations are leaving the built environment education sector in a quandary, with mixed signals and directions to go forward. Additionally, the lack of consistency is hampering coherence of change across the built environment education sector, reflecting exactly what is transpiring inside each institute. Thus, some courses are struggling to meet changes in accreditation expectations in an arena where management and numerical dominance by other disciplines sees little merit in such compliance nor the relevance in investing university time into scaffolding unit or subject renovations or enabling outreach ventures. On the one hand some academics want to change, but they do not know how to change and to enable change. On the other hand, students are saying that they want Indigenous learning and IKS, but that there is not the academic capacity to facilitate such, and that their institute is too conservative in addressing this topic (Jones et al. 2017; Tucker et al. 2016). Key issues are the perceived lack of exemplars and a lack of Indigenous people willing to engage with the tertiary sector and to become an academic in these disciplines.

The *Recasting Terra Nullius* project documented the abovementioned patterns and put forth a Guide to navigating through many of these issues (Jones et al. 2018). Tables 4.2, 4.3, and 4.4 provide project exemplars identified during this research (Jones et al. 2018). In the last 2 years AILA

Table 4.2 Key Australian architecture projects

State/Territory	Architecture project
Redfern, NSW	Aboriginal Dance Theatre
Weipa, Qld	Achimbun Interpretive and Visitor Information Centre
University of Newcastle, Newcastle, NSW	Birabahn Aboriginal & Torres Strait Islander Centre
Kakadu National Park, NT	Bowali Visitor Centre
Halls Gap / *Budja Budja, Gariwerd* / Grampians National Park, Vic	Brambuk Cultural Centre
Wugularr or Beswick, NT	Djakanimba Pavilions
Bathurst, NSW	Girrawaa Creative Works Centre
Pilbara, WA	Karijini National Park Visitors Centre
Edith Cowan University, Perth, WA	Kurongkurl Katitjin Centre for Indigenous Australian Education and Research
Yirrkala, NT	Marika Alderton House
Tennant Creek, NT	Nyinkka-Nyunyu Art and Culture Centre
Port Augusta, SA	Port Augusta Courts Complex
Warburton, WA	Tjulyuru Ngaanyatjarri Cultural and Civic Centre
Uluru, NT	Uluru-Kata Tjuta Cultural Centre
Derby, WA	West Kimberley Regional Prison
Wilcannia, NSW	Wilcannnia Health Service

Source: Jones et al. (2018: 34)

Table 4.3 Key Australian landscape architecture projects

State/Territory	Landscape architecture project
Adelaide, SA	Victoria Square / *Tarndanyangga*
Melbourne, Vic	Forest Gallery, Museum Victoria
Melbourne, Vic	Birrarung Marr, River Yarra
Melbourne, Vic	Bunjilaka, Museum Victoria
Broome, WA	*Walyjala-jala buru jayida jarringgun buru Nyamba Yawuru ngan-ga mirli mirli* (YRNTBC 2011)
South-western Victoria	*Ngootyoong Gunditj Ngootyoong Mara South West Management Plan* (Parks Victoria 2015)

Source: Jones et al. (2018: 35)

Table 4.4   Key Australian urban/regional planning projects

State/Territory	Planning project
Australia	*Planning for Country: Cross-cultural decision-making on Aboriginal lands* (Walsh and Mitchell 2002)
Bunya Bunya Mountains, Qld	*Bonye Bu'rū Booburrgan Ngmmunge: Bunya Mountains Aboriginal Aspirations and Caring for Country Plan* (BMECBMRG 2010)
Broome, WA	*Walyjala-jala buru jayida jarringgun buru Nyamba Yawuru ngan-ga mirli mirli: Planning for the future: Yawuru Cultural Management Plan* (YRNTBC 2011)
South-western Victoria	*Kooyang Sea Country Plan* (FAT and WMAC 2004),
South-western Victoria	*Ngootyoong Gunditj Ngootyoong Mara South West Management Plan* (Parks Victoria 2015)

Source: Jones et al. (2018: 36)

has issued its own 'Country Policy' statement (AILA 2019a), tabulating 20 relevant case studies (of which the *Recasting Terra Nullius* project was one, and the Gathering and the foreshadowed book was another (AILA 2019b)), as evidenced in Table 4.5, with their Victoria Group issuing an Award of Excellence to this project.

Citations and peer praise aside, policy change and application in education inside and outside universities in the Australian built environment sector pertinent to Indigenous Knowledge Systems is and has been slow. While the *Recasting Terra Nullius* project has been instrumental, it has been occurring inside differing professional institute policy change machinations, and even slower internal university policy and execution changes despite positive rhetoric from both parties. Therefore, it is but a small voice in a larger tangled educational web. The strength of voice for change is variable across the academic, student, practitioner and university bureaucratic sectors, being little heard, and little hearing the increasing student voice that is saying that the topic realm is relevant to their careers. Thus, the latter is falling on deaf ears and fragmented eyes, who are more caught up with dealing with maintaining predominant societal norms of economic efficient, educational policy and research quota imperatives.

Table 4.5 Position statement: Connection to country—case studies

State/Territory	Coded	Landscape architecture project
Uluru, NT	1	Uluru Kata-Tjuta National Park Cultural Centre
Mapoon, Qld	2	Old Mapoon Settlement Concept Plan
'… at the confluence of the Maribyrnong River and Arundal (Dry) Creek', Keilor, Victoria	3	Murrup Tamboore ('Spirit's waterhole') (formerly The Keilor Archaeological Site),
University of Tasmania, Launceston, Tas	4	Riawunna Aboriginal Studies Centre Landscape
Townsville, Qld	5	The Jezzine Barracks / Kissing Point / Garabarra
Mabuiag (Gumu) Island, Torres Strait, Qld	6	Mabuiag (Gumu) Island Village
Victoria Square, Adelaide, SA	7	Tarntanyangga / Victoria Square Regeneration Project
Rottnest Island, WA	8	Rottnest Island Coastal Walk Trails (Stage 1)
Little Bay, Sydney, NSW	9	Little Bay Cove
Port Adelaide, SA	10	Lartelare Aboriginal Heritage Park
Yawuru Country Broome and West Kimberley WA.	11	*Walyjala-jala burujayida jarringgun buru Nyamba Yauwuru ngan-ga mirli mirli— Planning for the future: Yawuru Cultural Management Plan* (YCMP) (YRNTBC 2011)
Perth, WA	12	Bilya Kard Boodja Lookout
Murujuga (Burrup Peninsula), WA.	13	*Ngaayintharri Gumawarni Ngurrangga: We all come together on this Country—Murujuga Cultural Management Plan* (MCMP) (2016)
Roebourne, WA	14	Roebourne Town Visioning
	15	*Recasting Terra Nullius Blindness: Empowering Indigenous Protocols and Knowledge in Australian University Built Environment Education* (Jones et al. 2017)
Tamborine Mountain, Qld	16	Main Street Village Green
Broome, WA	17	Jetty to Jetty
Melbourne, Vic	18	Everyone's Knowledge in Country: Yurlendj-nganjin
Queens Park, Perth, WA	19	Sister Kate's Home Kids Bush Block -Development Visioning
Perth, WA	20	Yagan Square,

Source: AILA (2019b)

REFERENCES

Architects Accreditation Council of Australia (AACA). (2020). *National Standard of Competency for Architects.* Available at: http://competencystandardforarchitects.aaca.org.au/matrix/index/print/assessment/1?assessment%5B%5D=1. Accessed 27 May 2020.

Australia. (1993). *Native Title Act 1993* (Cth).

Australian Capital Territory. (2004). *Architects Act 2004* (ACT).

Australian Institute of Architects Victoria (AIA). (2010a). Architects Accreditation Council of Australia and the Australian Institute of Architects (ACCA/AIA) 2012, Australian and New Zealand Architecture Program Accreditation Procedure. Accessed at: www.aaca.org.au. Accessed 27 May 2020.

Australian Institute of Architects Victoria (AIA). (2010b). *Reconciliation Action Plan.* Accessed at: www.raia.com.au. Accessed 1 Dec 2016.

Australian Institute of Landscape Architects (AILA). (2016). *Education Standards and Procedures (Endorsed Policy May 2015, V3 Revised 2 July 2016).* Australian Institute of Landscape Architects, Canberra. Accessed at: www.aila.org.au. Accessed 1 Dec 2016.

Australian Institute of Landscape Architects (AILA). (2019a). *Position Statement: Connection to Country.* Australian Institute of Landscape Architects, Canberra. Accessed at: www.aila.org.au. Accessed 1 May 2020.

Australian Institute of Landscape Architects (AILA). (2019b). Connection to *Country: Case Studies.* Australian Institute of Landscape Architects, Canberra. Accessed at: www.aila.org.au. Accessed 1 May 2020.

Australian Institute of Landscape Architects (AILA). (2019c). *Constitution.* Australian Institute of Landscape Architects, Canberra. Accessed at: www.aila.org.au. Accessed 1 May 2020.

Brunner, J., & Glasson, J. (2015). *Contemporary Issues in Australian Urban and Regional Planning.* London: Routledge.

Bunya Mountains Elders Council and Burnett Mary Regional Group (BMECBMRG). (2010). *Bonye Bu'rū Booburrgan Ngmmunge: Bunya Mountains Aboriginal Aspirations and Caring for Country Plan.* Canungra: Markwell Consulting & Bunya Mountains Elders Council and Burnett Mary Regional Group.

Byrne, J., Dodson, J., & Sipe, N. (2014). *Australian Environmental Planning: Challenges and Future Prospects.* London: Routledge.

Cooke, C. (2018). *Indigenous Fluency: In the Victorian Planning Profession.* Geelong: Unpublished MPlan Thesis, Deakin University.

Fantin, S. (2003). Aboriginal Identities in Architecture. *Architecture Australia (Architecture Media), 92*(5), 84–87.

Framlingham Aboriginal Trust (FAT) and Winda Mara Aboriginal Corporation (WMAC). (2004). *Kooyang Seu Country Plan*. Warrnambool: Aboriginal Trust and the Winda Mara Aboriginal Corporation.

Go-Sam, C. (2008). *Indigenous Design Paradigms: Working With and Against Indigenous Design Paradigms*. Architecture Australia.

Gurran, N., & Phibbs, P. (2003). Reconciling Indigenous and Non-Indigenous Land Management Concepts in Planning Curricula. *Proceedings of Australian and New Zealand Association of Planning Schools (ANZAPS) Conference*, Auckland, 26–28 September.

Gurran, N., & Phibbs, P. (2004). Indigenous Interests and Planning Education: Models from Australia, New Zealand and North America. *Australian and New Zealand Association of Planning Schools (ANZAPS) Conference 2004*. Association Historique Internationale de l'Ocan Indien.

Heyes, S., & Tuiteci, S. (2013). *Transects: Windows into Boandik Country*. Canberra: Faculty of Arts and Design, University of Canberra.

Jones, D. S. (2002). Time, Seasonality and Design. *Paper Presented at Australian Institute of Landscape Architects National Conference: People + Place – The Spirit of Design*, Darwin, 22nd–24th August 2002.

Jones, D. S., Low Choy, D., Revell, G., Heyes, S., Tucker, R., & Bird, S. (2017). *Re-casting Terra Nullius Blindness: Empowering Indigenous Protocols and Knowledge in Australian University Built Environment Education*. Canberra: Office for Learning and Teaching / Commonwealth Department of Education and Training. At http://www.olt.gov.au/project-re-casting-terra-nullius-blindness-empowering-indigenous-protocols-and-knowledge-australian- (2017–2018); at https://ltr.edu.au/resources/ID12_2418_Jones_Report_2016.pdf (2018). Accessed 1 May 2020.

Jones, D. S., Low Choy, D., Tucker, R., Heyes, S., Revell, G., & Bird, S. (2018). *Indigenous Knowledge in the Built Environment: A Guide for Tertiary Educators*. Canberra: Office for Learning and Teaching / Commonwealth Department of Education and Training. https://ltr.edu.au/resources/ID12-2418_Deakin_Jones_2018_Guide.pdf. Accessed 1 May 2020.

Lawson, G., & Erickson, M. (2002). Connecting Theory, Practice and Places in Today's Landscape: An Education Initiative. *Paper Presented at Australian Institute of Landscape Architects National Conference: People + Place – The Spirit of Design*, Darwin, 22nd-24th August 2002.

Lochert, M. (1997). Mediating Aboriginal Architecture: Collaborations Between Aboriginal Clients and Non-Aboriginal Architects. *Transition, 54-55*, 8–19.

Low Choy, D., & Van der Valk, A. (2011). Inclusive Engagement: Informing Planning Education for Indigenous Engagement in Regional Planning. *Proceedings of the 3rd World Planning Schools Congress*, 4–8 July 2011, Perth, UWA.

Low Choy, D., Wadsworth, J., & Burns, D. (2010). Seeing the Landscape Through New Eyes. *Australian Planner, 47,* 178–190.

Low Choy, D., Wadsworth, J., & Burns, D. (2011a). Looking after Country: Landscape Planning Across Cultures. *Paper Presented at Australian Institute of Landscape Architects National Conference: Transform, Brisbane,* 11th–13th August 2011.

Low Choy, D., & Van der Valk, A. (2011b), 'Inclusive Engagement: Informing planning education for Indigenous engagement in regional planning', Proceedings of the 3rd World Planning Schools Congress, 4–8 July 2011. UWA: Perth, WA.

Mallie, T., & Ostwald, M. (2009). Aboriginal Architecture: Merging Concepts from Architecture and Aboriginal Studies. In *Cumulus 38° South 2009 Conference. Cumulus 38°s 2009: Hemispheric Shifts Across Learning, Teaching and Research: Proceedings of the Cumulus Conference (Melbourne 12–14 November, 2009).* Melbourne: Swinburne University of Technology / RMIT University.

Memmott, P. C. (1997). Aboriginal Signs and Architectural Meanings. Part 2: Generating Architectural Signs. Architectural Semiotics in Designing for Aboriginal Clients. *Architecture Theory Review, 2*(1), 38–64.

Memmott, P. C. (2007). *Gunyah, Goondie + Wurley: The Aboriginal Architecture of Australia.* St Lucia: University of Queensland Press.

Memmott, P. C., & Chambers, C. D. (Eds.). (2003). *Take 2: Housing Design in Indigenous Australia.* Canberra: Royal Australian Institute of Architects.

Memmott, P. C., & Reser, J. (2000). Design Concepts and Processes for Public Aboriginal Architecture. In *PaPER - 11th Conference on People Physical Environment Research, University of Sydney, NSW, 3–6 December, 1998* (pp. 69–86). Sydney: PAPER.

Murujuga Aboriginal Corporation (MAC). (2016). *Ngaayintharri Gumawarni Ngurrangga: We all come together on this Country – Murujuga Cultural Management Plan.* Karratha, WA: Murujuga Aboriginal Corporation.

New South Wales. (2003). *Architects Act 2003* (NSW).

Nganampa Health Council Inc, South Australian Health Commission, Aboriginal Health Organisation of South Australia. (1987). *Report of Uwankara Palyanku Kanyintjaku: An Environmental and Public Health Review Within the Anangu Pitjantjatjara Lands.* Alice Springs: Nganampa Health Council.

Northern Territory. (1963). *Architects Act 1963* (NT).

Oberklaid, S. (2008). *Indigenous Issues and Planning Education in Australia.* Unpublished Master of Planning thesis, The University of Melbourne, Melbourne.

Parks Victoria, Gunditj Mirring Traditional Owners Aboriginal Corporation (GMTOAC) and Winda Mara Aboriginal Corporation (WMAC). (2015). *Ngootyoong Gunditj Ngootyoong Mara South West Management Plan.*

Melbourne: PV. http://parkweb.vic.gov.au/explore/parks/mount-eccles-national-park/plans-and-projects/ngootyoong-gunditj-ngootoong-mara. Accessed at 1 June 2020.

Pholeros, P. (2003). Housing for Health, or Designing to Get Water in and Shit Out. In P. Memmott (Ed.), *Take 2: Housing Design in Indigenous Australia*. Canberra: RAIA Monograph.

Planning Institute of Australia (PIA). (2016). Accreditation Policy for the Recognition of Australian Planning Qualifications. Accessed at: www.planning.org.au. Accessed at 1 Dec 2–16.

Planning Institute of Australia (PIA). (2019). *PIA Re-affirms Support for Indigenous Recognition Based on Uluru Statement from the Heart*. Available at: https://www.planning.org.au/news-archive/2018-2019-media-releases/pia-re-affirms-support-for-indigenous-recognition-based-on-uluru-statement-from-the-heart. Accessed 15 June 2020.

Planning Institute of Australia (PIA). (2020). *PIA's Support for the Uluru Statement Remains True*. PIA Member Email Issued 25 May 2020.

Porter, L. (2016). *Submission to the PIA Draft Accreditation Policy for the Recognition of Australian Planning Qualifications*. Unpublished Open Letter; Permission of the Author Granted to Use/Cite.

Queensland. (2002). *Architects Act 2002* (Qld).

Revell, G. (2002a). About the Role of Landscape Architecture in Rural/Regional Australia. *Kerb Journal of Landscape Architecture, 11*, 80–81.

Revell, G. (2002b). Ten Questions on Borree. *Kerb Journal of Landscape Architecture, 11*, 34–37.

Revell, G. (2014). An Educated Ecology of Wellbeing: Experimental Approaches to Education. *Landscape Architecture Australia, 139*, 28–29.

Sarkissian, W. (1996). *With Whole Heart: Nurturing an Ethic of Caring for Nature in the Education of Australian Planners*. Unpublished PhD Thesis, Murdoch University, Perth.

Sawyer, A. (2011). *Revealing Identity: How Symbolism Can be Used in Public Indigenous Architecture to Reflect Australian Indigenous Culture*. Unpublished Master of Architecture thesis, School of Architecture and Built Environment, Deakin University, Geelong.

Scally, S. (2003). Outstation Design: Lessons from Bawinanga Aboriginal Corporation in Arnhem Land. In P. Memmott (Ed.), *Housing Design in Indigenous Australia*. Red Hill: Royal Australian Institute of Architects.

Sinatra, J., & Murphy, P. (1999). *Listen to the People, Listen to the Land*. Carlton: Melbourne University Press.

South Australia. (2009). *Architectural Practice Act 2009* (SA).

Spurr, S. (2019, July 11). Reconfiguring Architecture's Relationship to Indigenous Knowledge: 2019 National Architecture Conference. *ArchitectureAU*. Available at: https://architectureau.com/articles/

reconfiguring-architectures-relationship-to-indigenous-knowledge/. Accessed 27 May 2020.

Stallard, K. E. (2011). *Where to Begin? An Exploration of How Architects Approach Aboriginal Housing in Australia.* Unpublished Master of Architecture Thesis, School of Architecture & Built Environment, Deakin University, Geelong.

Tasmania.(1929). *Architects Act 1929* (Tas).

Thompson, S., & Maginn, P. J. (Eds.). (2012). *Planning Australia: An Overview of Australian Urban and Regional Planning* (2nd ed.). Cambridge: Cambridge University Press.

Tonkinson, R. (2007). Aboriginal "Difference" and "Autonomy" Then and Now: Four Decades of Change in a Western Desert Society. *Anthropological Forum, 17*(1), 41–60.

Tucker, R., Low Choy, D., Heyes, S., Revell, G., & Jones, D. S. (2016). Re-Casting Terra Nullius Design-Blindness: Better Teaching of Indigenous Knowledge and Protocols in Australian Architecture Education. *International Journal of Technology and Design Education, 28*(1), 303–322.

Victoria. (1991). *Architects Act 1991* (Vic).

Walsh, F., & Mitchell, P. (Eds.). (2002). *Planning for Country: Cross-Cultural Approaches to Decision-Making on Aboriginal Lands.* Alice Springs: IAD Press.

Ward, M. (2011). *It's Not About Buildings: An Interview with Health Habitat's Paul Pholeros.* Architectural Review Australia.

Wensing, E. (2011). Improving Planners' Understanding of Aboriginal and Torres Strait Islander Australians and Reforming Planning Education in Australia. *Paper presented to the Third World Planning Schools Congress,* Perth, 4–8 July 2011.

Wensing, E. (2016). *'Going Deeper Than the Rhetoric': Comments on the PIA Draft Accreditation Policy for the Recognition of Australian Planning Qualifications.* Unpublished Open Letter, Permission of the Author Granted to Use/Cite.

Wensing, E., & Small, G. (2012). A Just Accommodation of Customary Land Rights. *Paper Presented to the 10th International Urban Planning and Environment Association Symposium,* University of Sydney, 24–27 July 2012.

Western Australia. (2004). *Architects Act 2004* (WA).

Williams, P., & Houston, S. (1997). Environmental Health Conditions in Remote and Rural Aboriginal Communities in Western Australia. *ANZ Journal of Public Health, 21*(5), 511–518.

Yawuru Registered Native Title Body Corporate (YRNTBC). (2011). *Yawurru Cultural Management Plan: Walyjala-jala buru jayida jarringgun buru Nyamba Yawuru ngan-ga mirli mirli.* Broome, WA: UDLA.

CHAPTER 5

Learning Environments and Contexts

David S. Jones, Kate Alder, Shivani Bhatnagar,
Christine Cooke, Jennifer Dearnaley, Marcelo Diaz,
Hitomi Iida, Anjali Madhavan Nair, Shay-lish McMahon,
Mandy Nicholson, Gavin Pocock, Uncle Bryon Powell,
Gareth Powell, Sayali G. Rahurkar, Susan Ryan,
Nitika Sharma, Yang Su, Saurabh V. Wagh,
and Oshadi L. Yapa Appuhamillage

D. S. Jones (✉) • U. B. Powell
Wadawurrung Traditional Owners Aboriginal Corporation,
Geelong, VIC, Australia
e-mail: davidsjones2020@gmail.com

K. Alder
Maribyrnong City Council, Melbourne, VIC, Australia

S. Bhatnagar • S. V. Wagh
Moir Landscape Architecture, Islington, NSW, Australia

C. Cooke • S. Ryan
School of Architecture & Built Environment, Deakin University,
Geelong, VIC, Australia
e-mail: ccooke@deakin.edu.au; rsusa@deakin.edu.au

© The Author(s), under exclusive license to Springer Nature
Singapore Pte Ltd. 2021
D. S. Jones (ed.), *Learning Country in Landscape Architecture*,
https://doi.org/10.1007/978-981-15-8876-1_5

61

5.1   Deakin University as a Learning Venue

Deakin University is a public university in Australia established in 1974 under the *Deakin University Act 1974* (Victoria 1974), and was named after the second Prime Minister of Australia, Alfred Deakin. The revised *Deakin University Act 2009* (Victoria 2009) includes a statement about

J. Dearnaley
Balyang Consulting, Geelong, VIC, Australia

M. Diaz
MINT Pool and Landscape Design, Melbourne, VIC, Australia
e-mail: marcediaz9940@gmail.com

H. Iida
Kihara Landscapes, Melbourne, VIC, Australia

A. M. Nair
Ground Ink, Sydney, NSW, Australia

S.-l. McMahon
GHD Woodhead, Melbourne, VIC, Australia
e-mail: Shay.McMahon@ghd.com

M. Nicholson
Tharangalk Art, Melbourne, VIC, Australia

G. Pocock
Garden Consultants, Geelong, VIC, Australia

G. Powell
Wadawurrung Traditional Owners Aboriginal Corporation,
Geelong, VIC, Australia

Legals Lawyers and Barristers (LLB) Pty Ltd, Canberra, ACT, Australia

S. G. Rahurkar
Inspiring Place, Hobart, TAS, Australia
e-mail: sayalirahurkar@gmail.com

N. Sharma
Mexted Rimmer Landscape Architecture, Geelong, VIC, Australia

Y. Su
Landscape Architect, Melbourne, VIC, Australia

O. L. Yapa Appuhamillage
Thomson Hay Landscape Architecture, Ballarat East, VIC, Australia

the Objects for which this university is required to ensure. An explicit statement about Indigenous culture in Clause 5 (f) states:

5 (f) to use its expertise and resources to involve Aboriginal and Torres Strait Islander people of Australia in its teaching, learning, research and advancement of knowledge activities and thereby contribute to—

(i) realising Aboriginal and Torres Strait Islander aspirations; and
(ii) the safeguarding of the ancient and rich Aboriginal and Torres Strait Islander cultural heritage;

In 2016, Deakin set goals to:

- *Deliver explicit Aboriginal and Torres Strait Islander inclusive curriculum to grow enrolments and completion rates and make this curriculum explicit in the Deakin Graduate Outcomes.*
- *All courses have embedded Aboriginal and Torres Strait Islander references and perspectives as part of the core units and appropriate electives.*
- *All degrees which have explicit Australian professional registration address Aboriginal and Torres Strait Islander contexts and perspectives as the essential starting point for understanding the Australian community and context.* (Paradies 2019: 1)

Deakin's Graduate Learning Outcomes (GLOs) describe the knowledge and capabilities graduates have acquired and are able to apply and demonstrate at the completion of their course (Deakin 2020). They consist of outcomes specific to a particular discipline or profession as well as transferable generic outcomes that all graduates should have acquired irrespective of their discipline area. Learning outcomes are not confined to the knowledge and skills acquired within a course, but also incorporate those that students bring with them upon entry to the course consistent with the Australian Qualifications Framework pathways policy (AQFC 2013). Deakin's courses are designed to ensure that students develop systematic knowledge and understanding of their discipline or chosen profession appropriate to their level of study. They are specified at the course level, mapped to course components and are assessed. In professionally accredited courses, discipline-specific learning outcomes may be defined in part by the relevant professional body.

Table 5.1 sets out these eight GLOs.

Reflecting upon these goals, a raft of policy rhetoric statements around Australia, and what actions other universities have taken in articulating and incorporating this topic into their GLOs (see Table 5.2), in late 2019

Table 5.1 Deakin University's Graduate Learning Outcomes (GLOs)

GLO1 Discipline knowledge and capabilities: appropriate to the level of study related to a discipline or profession

GLO2 Communication: Using oral, written and interpersonal communication to inform, motivate and effect change

GLO3 Digital literacy: Using technologies to find, use and disseminate information

GLO4 Critical thinking: Evaluating information using critical and analytical thinking and judgment

GLO5 Problem solving: Creating solutions to authentic (real world and ill-defined) problems

GLO6 Self-management: Working and learning independently, and taking responsibility for personal actions

GLO7 Teamwork: Working and learning with others from different disciplines and backgrounds

GLO8 Global citizenship: Engaging ethically and productively in the professional context and with diverse communities and cultures in a global context

Deakin (2020)

Table 5.2 Select Australian university Graduate attributes (or GLOs) focused on Aboriginal and Torres Strait Islander peoples

Curtin University	Culturally competent to engage respectfully with local First Peoples and other diverse cultures: Graduates will demonstrate cross-cultural capability and have an applied understanding of local First Peoples' "katajininy warniny" (translated from the Nyungar language as "ways of being, knowing and doing").
Griffith University	Culturally capable when working with First Australians
RMIT University	Show understanding of the social and cultural heritage of Aboriginal and Torres Strait Islander peoples in Australia through active engagement with individuals and communities. Analyse and examine issues of social justice and equality with respect to Aboriginal and Torres Strait Islander people and individuals.
The University of Melbourne	An understanding of and deep respect for Indigenous knowledge, culture and values.
The University of Sydney	Celebrates Aboriginal and Torres Strait Islander cultures, knowledge systems, and a mature understanding of contemporary issues.
The University of Western Australia	Respect Indigenous knowledge, values and culture.

Source: Paradies (2019: 4)

Deakin's Academic Board approved the amendment of GLO 8 to read: 'GLO8 Global citizenship and Indigenous engagement: Engaging ethically and productively in the professional context, and with Aboriginal and Torres Strait Islander peoples as well as diverse communities and cultures in a global context', effective the end of 2021.

This policy change recognizes that 'over the last three years, Deakin has introduced two generalist [undergraduate] units of study focused on Aboriginal and Torres Strait Islander knowledges … that are currently available for students in all courses at the University as electives. However, elective units do not ensure that all students study Aboriginal and Torres Strait Islander knowledges' (Paradies 2019: 1). No appraisal was made of extant relevant units elsewhere in Deakin, accordingly SRL733 was not identified at the time.

5.2   SRL733 INDIGENOUS MEANINGS AND PROCESSES

The MLArch degree, a 2-year equivalent AQF9 qualification, was promoted by changes in staff profiles and in the School of Architecture and Built Environment at Deakin. The market assessment identified the need for course distinctiveness, resulting in a rich blend of cross-cultural place-making and design with a major online learning route. Such a degree continues to cater for a student catchment uninterested in mainstream Australian landscape architecture educational experiences and learning modes that offer staff with an international flavour in outlook and research activities, demonstrating its economic viability and market catchment niche.

The origins of SRL733 Indigenous Meanings and Processes (SRL733) lie in newly arrived university staff experiences and values, rather than existing academic staff aspirations. Internally, the formative structure of the SRL733 was guided in development over 2011–2012 by the thoughts and discussions with two senior Aboriginal academics inside Deakin, and a well-respected Elder in the Geelong (*Djillong*) region. In 'designing' the unit, advice and direction was directly sought from academic scholars, Professor Wendy Brabham (of *Wamba Wamba, Wergaia, Nyeri Nyeri* and *Dhudhuroa* descent), *Gunditjmara* man Professor Mark Rose (2017), *Arabana* woman Professor Veronica Arbon (2008a, b), then at Deakin. Additional guidance was sought from the Geelong-based Wathaurong

Aboriginal Co-Operative's (WAC) Senior Cultural *Tandop* (Uncle) David Tournier Sr (of *Wemba Wemba*, *Yorta Yorta* and *Ngarrindjeri* ancestry) (Powell et al. 2019).

All persons were very appreciative of being consulted, and even being simply asked to advise. Most were surprised at being asked about the active decolonisation learning strategy proposed to be employed, the wide First Nations ambit and not just an Aboriginal focus, as well as the use of a non-traditional learning outcomes strategy that positioned learning in *Country* itself rather than objectifying the 'Aboriginal'. *Tandop* Tournier eagerly offered to actually participate in SRL733's teachings. Additional thoughts came from *Gunditjmara* man Damein Bell (Bell and Jones 2011), *Boon Wurrung* Elder *N'Arweet* (Aunty) Carolyn Briggs (2008), and *Kaurna* Elder Uncle Lewis O'Brien (2007), all three of which have maintained some minor oversight over SRL733. Thus, active Indigenous consultation in the unit's design, its aims and learning outcomes, was employed in line with research expectations about 'prior informed consent', embodied in the Australian Institute of Aboriginal and Torres Strait Islander Studies' (AIATSIS) *Guidelines for Ethical Research in Australian Indigenous Studies* (2002, 2012) and approvals to proceed secured.

Such discussions and consultations occurred over the course of 2010–2017, while the legal and cultural configurations of the Aboriginal cultural heritage knowledge in the *Djillong* (Geelong) region changed from the social infrastructure provider WAC to the Wadawurrung (Wathaurung Aboriginal Corporation) (W(WAC)), given the latter's recognition as the official Registered Aboriginal Party of under the Victorian *Aboriginal Heritage Act 2006* (Victoria 2006). A co-signed Reconciliation Action Policy with the City of Greater Geelong Council formalised this transition (COGG 2014). Additionally, *Tandop* David passed into the Dreaming in late 2016. In advance of this transition, Elders representatives of the W(WAC) were invited into the teaching of SRL733, an invitation eagerly accepted to participate as Wadawurrung people, and eagerly continue this association today as noted in this article's co-authorship.

Additionally, from 2015, Aboriginal people from *Wadawurrung*, *Eora*, *Awabakal*, *Gunai/Kurnai*, *Gunditjmara*, *Wurundjeri*, *Yorta Yorta* descent have been regular or intermittently involved in teaching into SRL733.

Interestingly, in the 2012 period, the nature of relevant readings about this topic, and linked to the built environment professions, was exceedingly limited. While this has changed considerably by 2020, the core references at this time that were used as teaching texts talked of *Country* (Clarke 1996, 2005, 2007, 2009; Jones 1993; Massola 1968; Rose 1992,

1996), the biogeographical evolution of Australia (Flannery 1994), *Gunditjmara* peoples (2010) and their *Country* (Dawson 1881), *Gagudju Country* (Neidjie et al. 1985), *Yawurru* peoples and their *Country* (Benterrak et al. 1996), *Adnyamathanha Country* (Tunbridge 1988), *Arrente Country* (Breeden 1997), bush foods or 'bush tukka' (Isaacs 2002), ethnobotany (Clarke 2007; Gott 1983; Gott and Conran 1991; Zola and Gott 1992), seasons and 'calendars' (Jones et al. 1997, 1998; Heyes 1999), *Gagudju Country* (Neidjie et al. 1985), archaeoastronomy (Clarke 2003, 2009, 2014), and classic papers by authors including Flood (1996), Rose (1988, 1996), McBryde (1978), Lourandos (1983), Morieson (1966), Gerritsen (2000), and Basso (1984).

The unit passed quietly through the Deakin University administrative academic approval procedures in late 2011, within the overall MLArch degree, and was first offered as a course and unit in Trimester 1 2012. Coincidentally, the MLArch received professional accreditation from AILA in late 2011, with a notation that SRL733 was well received.

SRL733 is drafted to respond to Deakin's GLOs and the lower-level Unit Learning Outcomes (ULOs). Learning activities and the four assignments in SRL733 are mapped specifically against the ULOs. Historically, the four assignments have remained relatively constant, geographically neutral, place-neutral, *Country*-neutral, enabling both on-campus and online students to self-select places and/or communities themselves, and not be constrained by assignment requirements that are *Wadawurrung Country*-specific.

The only student-led bias in assignment responses has been that many students have been fascinated in one assignment by the iconic and multi-awarded Greg Burgess & Associates-designed Uluru-Kata Tjuta Cultural Centre at *Uluru* on *Arrernte Country* (Findley 2002; Tawa 1996) and the Brambuk Cultural Centre in *Gariwerd*/The Grampians (Dovey 1996) because of their encapsulating photographs, but this has not been seen as a negative in their learning inquiries. Thus, assignment content responses have drawn content from India, Japan, Canada, New Zealand / *Aotearoa*, Malaysia, Vietnam, Sri Lanka and non-*Wadawurrung* places/*Country*'s and Indigenous communities and not from specific Aboriginal and Torres Strait Islander people and places that one would normally pre-suppose SRL733 assignments would attract. This has resulted in an international collage of assignment topics being submitted, discussed, read, and a politically safe context for international students to discuss Indigenous issues of their home nation, and for Aboriginal students to either openly talk about their ancestry and culture or to quietly raise topics whilst not identifying themselves as having Aboriginal ancestry (Figs. 5.1, 5.2, 5.3, 5.4, 5.5, 5.6, 5.7, 5.8 and 5.9).

Several aspects of the open classroom discussions have been evident, noting that the audience is post-graduate students.

Fig. 5.1 *barrangal dyara* (skin and bones) art installation, authored by Jonathan Jones, The Royal Botanic Gardens Sydney, NSW. (Image author: DS Jones 2016)

Fig. 5.2 Ballarat Indigenous Playground, designed by Jeavons Landscape Architects with Billy Blackall, Lake Wendouree, Ballarat, Vic. (Image author: DS Jones 2016)

The majority of Indian and Malaysian students have openly admitted that they did not know about, nor had been taught or been introduced to their home Indigenous peoples and their IKS.

For Malaysian students, people of the *Orang Asli* communities on Peninsula Malaysia were little heard of prior to this class.

For Indian students, the topic appears to be positioned in secondary school Indian history classes, where a mention of Scheduled Tribes occurred or through chance optional undergraduate design studios that ventured into Scheduled Tribe villages and landscapes. Scheduled Castes (SCs) and Scheduled Tribes (STs) are officially designated groups of historically identified peoples in India who comprise about 16.6% and 8.6%, respectively, of India's population (according to the 2011 census). The Indian *Constitution (Scheduled Castes) Order, 1950* (India 1950a) lists

Fig. 5.3 Barak Building, designed by ARM Architecture featuring William Barak's face (or the face of Traditional *ngurungaeta* (Elder) William Beruk of the *Wurundjeri-Wilam*), Melbourne, Vic. (Image author: DS Jones, 2018)

Fig. 5.4 Barak Building looking from the Shrine plaza, designed by ARM Architecture featuring William Barak's face (or the face of Traditional *ngurun-gaeta* (Elder) William Beruk of the *Wurundjeri-Wilam*), Melbourne, Vic. (Image author: DS Jones, 2018)

1108 castes across 29 states in its First Schedule, and the *Constitution (Scheduled Tribes) Order, 1950* (India 1950b) lists 744 tribes across 22 states in its First Schedule. Indian students do, however, bring to SRL733 a strong understanding of their own nation's historical experience with cultural hegemony through waves of invasion and occupation, Mughal followed by British, through to eventual independence and self-rule (Sanyal 2012; Thadoor 2015, 2017).

For both Chinese and Malaysian students, Indigenous peoples are totally absent in their secondary and tertiary schooling. As one Chinese student (participant SRR013) has observed, 'In the design courses [in China], we are encouraged to integrate the culture and protocols of the minorities into the design, while it is not an integrated educational system yet, and as students, we are fixed at a really vague understanding, we also lack the connection of applying these elements into the design outcomes'.

Fig. 5.5 Birrarung Marr, master design by Taylor Cullity Lethlean (TCL) in collaboration with Paul Thompson and Swaney Draper, featured sculpture *'Birrarung Wilam'* designed by Treahna Hamm, Lee Darroch and Vicki Couzens in collaboration with Mandy Nicholson and Glenn Romanis, Melbourne, Vic. (Image author: DS Jones, 2014)

Fig. 5.6 Forest Gallery, Museum Victoria, Taylor Cullity Lethlean (TCL) in collaboration with Paul Thompson and Mark Stoner, Melbourne, Vic. (Image author: DS Jones, 2014)

For Indonesian and Vietnamese students this could be considered in the light of their recent governance amalgamations in creating their nations, and/or internal wars, citing recognition of the cultural and design differences. For example, in Buddhist-influenced *Bali Aga*, the Islamic conservative state of Aceh in northern Sumatra, the *Minahasan* on Sulawesi, the *Ainu* on Hokkaidō, and the *Degar* tribes of central Vietnam and the *Hmong* (*Miao*), *Zao* and *Tay* of northern Vietnam.

A clear undercurrent in the classes has been a willingness to openly critique the impact of European colonialism upon their respective culture, whether Indigenous or not. This includes the dismantling of colonialism, its 'positive' development transformative role (Thadoor 2017), the desire to rewrite knowledge that occurred under colonial occupation(s), a rethinking about the validity of non-Western seasonal calendars, a questioning about 'what is our identity post-colonialism', and thinking about

Fig. 5.7 Reconciliation Place, designed by Kringas Architecture in collaboration with Sharon Payne, Alan Vogt, Amy Leenders, Agi Calka and Cath Eliot, with various artists and authors (Judy Watson, Michael Hewes, Matilda House, Vic McGrath, Joseph Elu, Kevin O'Brien, Jennifer Marchant, Simon Kringas, Sharon Payne, Marcus Bree, Benita Tunks, Graham Scott-Bohanna, Andrew Smith, Karen Casey, Cate Riley, Darryl Cowrie, Archie Roach, Lou Bennett, Pep Gascoigne, Thancoupie Gloria Fletcher, Jerko Starcevic, Kev Carmody, Paul Kelly, Belinda Smith, Rob Tindal, Munnari John Hammond, Alice Mitchell Marrakorlorlo, Wenten Rubuntja, Mervyn Rubuntja, Bill Neidjie, Djerrkura family, Paddy Japaljarri Stewart, and Cia Flannery, Canberra, ACT. (Image author: DS Jones, 2019)

what are the hidden cultural narratives behind our home landscapes? The most common responses at the first class are: 'We have not been taught this', 'This is irrelevant to our discipline', 'I don't know anything about this' and 'Why is this not about Aboriginals?'

The pedagogical approach operates as day-long intensives, consisting of six one-day lecture/workshop sessions alternating fortnightly on the same day. The structure enables immersiveness and the use of a common

Fig. 5.8 Recreated Tyrendarra stone house, Tyrendarra Indigenous Protected Area, Gunditj Mirring Aboriginal Traditional Owners Corporation, Tyrendarra, Vic. (Image author: DS Jones, 2015)

day-theme in content along with plenty of time for self-reflection and assignment research. Such is in contrast to traditional post-graduate units that are typically 1–2 hours of lectures and 2–3 hours of seminars or design studios randomly timetabled across the week for the semester/trimester.

The academic content of SRL733, that considers contemporary engagements and relationships with Australian and international Indigenous communities in design and planning practice and projects, is fourfold in its consideration of:

- Indigenous peoples and their cultural and spiritual relationships to land, territory, *Country*, language, name, knowledge transferal, sedentary patterns, custodianship, curatorship, alternate approaches to 'natural science', personal and environmental health and their symbiotic use and curatorship of natural resources as legitimate land design, planning and management tools and approaches;

Fig. 5.9 Brambuk Living Cultural Centre, *Budja Budja* / Halls Gap, designed by Greg Burgess Architects (Greg Burgess, David Mayes, Deborah Fisher, Simon Harvey, Des Cullen, Peter Ryan, Anthony Capsalis) Gariwerd / The Grampians, Vic. (Image author: DS Jones 2018)

- a set of Australian and international exemplar case studies where Indigenous peoples have served either as client or consultant in the formulation of design and planning projects that have resulted in international and highly significant, innovative and creative outcomes that demonstrate respect and cultural richness;
- processes of managing cultural-rich projects including consultation, engagement and protocols; and,
- on-site engagements with a place rich in Indigenous meanings, associations, history, myth, and providing a first-hand understanding of Indigenous protocols.

SRL733s-stated Handbook-listed Aims and ULOs are set out in Tables 5.3 and 5.4. The text has been subject to internal review processes since 2012 but has remained relatively consistent in their language and intent

Table 5.3   SRL733s Aims and Learning Objectives

Aims The main aims of the unit are:	To introduce and instill an appreciation of Indigenous culture and their inter-relationships to land and the environment including concepts of territory, country, custodianship, season and time, language and name, myth and meaning; To appreciate the role of protocols in Indigenous culture; To consider international and Australian design and planning exemplars that resulted in Indigenous and culturally-rich significant, attuned and relevant outcomes To provide a sound understanding of inter-relationship of Indigenous cultures to natural systems, natural forces that shape and determine human responses, including curatorship responsibilities and activities that shape landscape and communities; To meet and address professional institutional policies and expectations that warrant a clear desire to foster an appreciation, respect and understanding of indigenous culture, its inter-relationship to land and environment, and processes of engagement to enrich design and planning projects.
Learning Objectives	To develop student ethic, understanding and respect of Indigenous communities, whether Australian or International, and the deep knowledge and skill base they have to their country; To introduce students to Indigenous concepts of land and ecological science, in particular seasons, time, plants, animals, land management, and the roles they might play in design and planning projects; To demonstrate and explore contemporary design and planning exemplars, that have and are continuing to closely work with or for Indigenous communities, that have and are resulting in culturally-rich outcomes and peer-acclamation for their innovation and cultural relevance; To develop practical skills appreciating and respecting Indigenous protocols, engagements on major design projects and thereupon during construction and management phases; To review and discuss professional practice and community ethics, policies and strategies as they relate to Indigenous engagement, respect and development creation.

since originally drafted and approved, although with additional amendments to encompass 'public health'.

Thematically, SRL733 considers and investigates, per session:

- The nature of Indigenous communities and knowledge systems, their unique relationships to lands/waters/*Country*'s, but also where and how built environmental practice engages today with such communities;

Table 5.4 SRL733s Unit Learning Outcomes (ULOs)

ULO:	At the completion of this unit students can:	The ULO relates to the following GLO/s:
ULO1	Articulate an ethical understanding of and respect for Indigenous communities, whether Australian or international, and the deep knowledge and skill base such communities have to their *Country* (or equivalent)	GLO1 GLO8
ULO2	Discuss and use Indigenous concepts of land and ecological science, in particular seasons, time, plants, animals, land management, environmental and personal health, and the roles they might play in design and planning projects	GLO1
ULO3	Critique contemporary design and planning exemplars, that have and are continuing to closely work with or for Indigenous communities, that have and are resulting in culturally rich outcomes and peer-acclamation for their innovation and cultural relevance	GLO4 GLO8
ULO4	Apply practical skills to appreciate and respect Indigenous protocols, engagements on major design projects and thereupon during construction and management phases	GLO2 GLO3 GLO4
ULO5	Critique and discuss professional practice and community ethics, policies and strategies as they relate to Indigenous engagement, respect and development creation.	GLO1 GLO4 GLO8

- Indigenous learning approaches and decolonisation theory;
- A condensed historical journey through indigenous communities, from **60,000** years ago onwards (e.g. Flannery 1994), with attention to Aboriginal and Torres Strait Islander peoples and India Scheduled Tribes, including environmental impacts and cultural pre- and post-colonialism and post-independence/federation contexts; the problems with colonised words and terms (such as '*terra nullius*', 'colonialized', 'settled', 'explorer', 'discovered', 'Scheduled Tribes', 'Aboriginal', '*Country*', 'Indigenous', etc. as well as concepts of 'time' as explaining when incidents occurred in the past);
- The way Indigenous people envisage Knowledge Systems, language and their environment including relationships to resources (flora/fauna/water/stone/etc.), familial systems and how such are linked to resource custodianship and language, time, seasons, myths and (historical) narratives, management of places, skies and waters;

- A deeper reading of the environment, focused upon knowing seasonality, temporality and language and the location/translation of this knowledge into plans and designs that are culturally responsive; and,
- A review of key landscape architect (architectural and urban planning) design/plan precedents across Australasia and southern Asia that are Indigenous-rich and have been devised to direct engagement with relevant Indigenous people(s) or community(ies) to appreciate the engagement process, the quality and Indigenous-aligned outcomes, irrespective of the peer design plaudits.

A facet in the teaching approach is the aspiration of positioning indigenous science on the same 'pedestal' as Western science, and similarly the design and planning disciplines. In this way, one is teaching and enticing students to appreciate and discern both cultures and their legacies and differences equally and respectfully, rather than assuming that all answers lie in western knowledge.

Such a learning strategy, and the educational journey of SRL733, has had a major influence upon MLArch students in their discussions and self-selected projects. In a School that is predominantly host to architecture and construction management students, it is the landscape and planning students who are commencing their oral assessable presentations in other units and in their capstone units with 'Acknowledgement of Country' statements (when involving Australian sites). Additionally, with supporting drawings and PowerPoint pages possessing clear indigenous site analysis content, selecting capstone projects that require a place/tradition of Indigenous values, raising interest from the postgraduate architecture, public health, urban planning, engineering sustainability and construction management students to consider selecting SRL733 as an elective. Of those architecture students who have done so, nearly all have undertaken capstone design projects that are Indigenous-rich.

5.3   INTERNAL LEARNING OUTCOMES EVALUATIONS

Like most Australia universities, there is an imperative to statistically and quantitatively evaluate, monitor and improve data that identifies learning and research 'production' by a university. The latter statistics are very much bound into national and international benchmarking, and is broken down in the sub-category discipline areas to compare discipline across universities. The latter unfortunately bundles all built environment courses into one 'Built Environment' subcategory, making it difficult to evaluate disciplines. This statistical approach also drives staff advancement and engagement activities.

Teaching, or rather 'learning outcomes', has its own statistical protocols of which evaluation of 'learning outcomes' at an on-ground level is of a lesser priority to research. At Deakin, the 'Student Evaluation of Teaching and Units' (SETU) is now superseded by eVALUate. Both are optional student e-surveys undertaken at the end of each unit's academic studies, before each student's results are released.

There is generically around Australia considerable academic teacher angst that such evaluation tools, optional and not mandatory, tend to reflect popularity polls, avenues for students to vent negativisms about content, marking, perceived relevance of the unit, and the actual teacher personality. The questions asked often are immediate in content, not linked to the nature of the course and unit itself or their CLOs or ULOs, and lack longevity of change of values of influence and self-reflectivity.

SRL733 has been subject to both SETU, that ceased in 2013, and eVALUate, that commenced in 2014, evaluation tools (Table 5.5).

In examining the eVALUate scores against the standard questions, there is a common consistency across 2015–2018 responses. Prior to 2015, SETU was applied, and it had a different set of questions that cannot be compared to this set. This data is obtained during a students' enrolment in the unit, in the last 2–3 weeks of their enrolment, and before final unit results are listed. This optional questioning occurs at the time when students are concerned about their assignment responses in this and other units they are enrolled in the same trimester, pending examinations, responsiveness of staff to their in-class and online questions, and thus elements of bias and frustration can influence results.

Table 5.5 SETU and eVALUate generic data for SRL733

Enrolment year	SETU / eVALUate #	Response rate %	On campus student #	OnCampus response rate %	Online student #	Online response rate %
2015	11	44	9	41	2	67
2016	24	52	32	53	14	50
2017	16	48	13	48	3	30
2018	30	47	24	50	6	38

Table 5.6 SETU and eVALUate specific data for SRL733

	2015			2016			2017			2018		
	Agree	Disagree	Unable to judge	Agree	Disagree	Unable to judge	Agree	Disagree	Unable to judge	Agree	Disagree	Unable to judge
The learning outcomes in this unit are clearly identified.	100	0	0	100	0	0	94	0	0	93	3	3
The learning experiences in this unit help me to achieve the learning outcomes.	100	0	0	100	0	0	100	0	0	90	7	3
The learning resources in this unit help me to achieve the learning outcomes.	82	18	0	100	0	0	94	0	0	93	3	3
The assessment tasks in this unit evaluate my achievement of the learning outcomes.	82	9	0	100	0	0	100	0	0	93	7	0

(continued)

Table 5.6 continued

	2015			2016			2017			2018		
	Agree	Disagree	Unable to judge	Agree	Disagree	Unable to judge	Agree	Disagree	Unable to judge	Agree	Disagree	Unable to judge
Feedback on my work in this unit helps me to achieve the learning outcomes.	82	18	0	96	4	0	100	0	0	93	7	0
The workload in this unit is appropriate to the achievement of the learning outcomes.	91	9	0	96	4	0	94	6	0	93	7	3
The quality of teaching in this unit helps me to achieve the learning outcomes.	100	0	0	96	4	0	93	0	7	97	3	0
I am motivated to achieve the learning outcomes in this unit.	100	0	0	96	4	0	100	0	0	97	3	0

I make best use of the learning experiences in this unit.	91	0	0	96	0	94	6	0	87	3	10
I think about how I can learn more effectively in this unit.	91	9	0	100	0	93	7	0	93	7	0
Overall, I am satisfied with this unit.	100	0	0	100	0	94	0	6	93	7	0

Statistical responses in this unit, in Table 5.6, demonstrate a high level of engagement, interest, and receptiveness to the content, its assignments, and its learning activities.

REFERENCES

Arbon, V. (2008a). Knowing from Where? In A. Gunstone (Ed.), *History, Politics and Knowledge: Essays in Australian Studies* (pp. 145–146). North Melbourne: Australian Scholarly Press.

Arbon, V. (2008b). *Arlathirnda Ngurkarnda Ityirnda: Being-Knowing-Doing: De-colonising Indigenous tertiary education*. Teneriffe: Post Pressed.

Australian Institute of Aboriginal and Torres Strait Islander Studies (AIATSIS). (2002). *Guidelines for Ethical Research in Australian Indigenous Studies*. Canberra: AIATSIS. Available at: https://aiatsis.gov.au/research/ethical-research/guidelines-ethical-research-australian-indigenous-studies. Accessed 1 May 2020.

Australian Institute of Aboriginal and Torres Strait Islander Studies (AIATSIS). (2012). *Guidelines for Ethical Research in Australian Indigenous Studies*. Canberra: AIATSIS. Available at: https://aiatsis.gov.au/research/ethical-research/guidelines-ethical-research-australian-indigenous-studies. Accessed 1 May 2020.

Australian Qualifications Framework Council (AQFC). (2013). *Australian Qualifications Framework: Second Edition, January 2013*. Adelaide: Australian Qualifications Framework Council. Accessed at: https://www.aqf.edu.au/sites/aqf/files/aqf-2nd-edition-january-2013.pdf. Accessed 1 May 2020.

Basso, K. H. (1984). 'Stalking with Stories': Names, Places, and Moral Narratives Among the Western Apache. In E. M. Bruner (Ed.), *Text, Play, and Story: The Construction and Reconstruction of Self and Society* (pp. 19–55). Prospect Heights: Waveland.

Bell, D., & Jones, D. S. (2011). Reinstating Indigenous Landscape Planning Curatorship: The Lake Condah Restoration Project. In *Proceedings of the 3rd World Planning Schools Congress, Planning's Future, 4–8 July 2011, Perth, WA*. Perth: UWA & ANZAPS.

Benterrak, K, S Muecke & P Roe with R Keogh, B(J) Nangan & EM Lohe. (1996). *Reading the Country: Introduction to Nomadology*. Fremantle: Fremantle Arts Centre Press.

Breeden, S. (1997). *Uluru: Looking after Uluru-Kata Tjuta the Anangu Way*. JB Books.

Briggs, C. (2008). *The Journey Cycles of the Boonwurrung Stories with Boonwurrung Language*. Melbourne: Victorian Aboriginal Corporation for Languages.

Clarke, P. A. (1996). Adelaide as an Aboriginal Landscape. In V. Chapman & P. Read (Eds.), *Terrible Hard Biscuits. A Reader in Aboriginal History: Journal of Aboriginal History* (pp. 69–93). Sydney: Allen & Unwin.

Clarke, P. A. (2003). Australian Aboriginal Mythology. In J. Parker & J. Stanton (Eds.), *Mythology. Myths, Legends, & Fantasies* (pp. 382–401). Sydney: Global Book Publishing.

Clarke, P. A. (2005). Aboriginal 'Fire-Stick' Burning Practices on the Adelaide Plains, pp. 424. In C. B. Daniels & C. J. Tait (Eds.), *Adelaide. Nature of a City. The Ecology of a Dynamic City from 1836 to 2036* (pp. 428–429). Adelaide: BioCity – Centre for Urban Habitats.

Clarke, P. A. (2007). *Aboriginal People and Their Lands*. Dural: Rosenberg Publishing Ltd.

Clarke, P. A. (2009). An overview of Australian Aboriginal ethnoastronomy. Archaeoastronomy. *The Journal of Astronomy in Culture. 21*, 39–58.

Clarke, P. A. (2014). Australian Aboriginal Astronomy and Cosmology. In C. L. N. Ruggles (Ed.), *Handbook of Archaeoastronomy and Ethnoastronomy*. New York: Springer.

Dawson, J. (1881). *Australian Aborigines: The Language and Customs of Several Tribes of Aborigines in the Western District of Victoria, Australia*. Melbourne: George Robertson (Facsimile ed. (1981), Canberra: Australian Institute of Aboriginal Studies).

Deakin University. (2020). *Graduate Learning Outcomes (GLOs)*. https://www.deakin.edu.au/about-deakin/teaching-and-learning/deakin-graduate-learning-outcomes. Accessed 1 June 2020.

Dovey, K. (1996, July). Architecture for the Aborigines. *Architecture Australia*. Available at: https://architectureau.com/articles/architecture-for-the-aborigines/. Accessed 1 June 2020.

Findley, L. (2002). Uluru-Kata Tjuta Cultural Centre Gregory Burgess. *Baumeister*, (3), 74–79.

Flannery, T. (1994). *The Future Eaters: An Ecological History of the Australasian Lands and People*. Sydney: New Holland Publishers.

Flood, J. (1996). *Moth Hunters of the Australian Capital Territory: Aboriginal Traditional Life in the Canberra Region*. Downer: JM Flood, c.1996.

Gerritsen, R. (2000). *The Traditional Settlement Patterns in South West Victoria Reconsidered*. Ballarat: R Gerritsen.

Gott, B. (1983). Murnong—*Microseris scapigera*: A Study of a Staple Food of Victorian Aborigines. *Australian Aboriginal Studies, 1*(2), 1–18.

Gott, B., & Conran, J. (1991). *Victorian Koorie Plants*. Hamilton: Yangennanock Women's Group, Aboriginal Keeping Place.

Greater Geelong, City of (COGG). (2014). *Karreenga Aboriginal Action Plan 2014–2017*. Geelong: City of Greater Geelong.

Heyes, S. A. (1999). *The Kaurna Calendar: Seasons of the Adelaide Plains.* Unpublished BLArch Hons thesis, University of Adelaide, Adelaide, SA.

India. (1950a). *Constitution (Scheduled Castes) Order, 1950.* Available at: https://www.constitutionofindia.net/constitution_of_india. Accessed 1 May 2020.

India. (1950b). *Constitution (Scheduled Tribes) Order, 1950.* Available at: https://www.constitutionofindia.net/constitution_of_india. Accessed 1 May 2020.

Isaacs, J. (2002). *Bush Food: Aboriginal Food and Herbal Medicine.* Sydney: New Holland Publishers.

Jones, D. S. (1993). *Traces in the Country of the White Cockatoo (Chinna junnak cha knæk grugidj): A Quest for Landscape Meaning in the Western District, Victoria, Australia.* Unpublished PhD Thesis, University of Pennsylvania, Philadelphia.

Jones, D. S., Low Choy, D. G., Revell, G., Heyes, S., Tucker, R., & Bird, S. (2017). *Re-casting Terra Nullius Blindness: Empowering Indigenous Protocols and Knowledge in Australian University Built Environment Education.* Canberra, ACT: Office for Learning and Teaching / Commonwealth Department of Education and Training. At http://www.olt.gov.au/project-re-casting-terra-nullius-blindness-empowering-indigenous-protocols-and-knowledge-australian-(2017–2018); at https://ltr.edu.au/resources/ID12_2418_Jones_Report_2016.pdf (2018), (accessed 1 May 2020).

Jones, D. S., Low Choy, D., Tucker, R., Heyes, S., Revell, G., & Bird, S. (2018). *Indigenous Knowledge in the Built Environment: A Guide for Tertiary Educators.* Canberra, ACT: Office for Learning and Teaching / Commonwealth Department of Education and Training; https://ltr.edu.au/resources/ID12-2418_Deakin_Jones_2018_Guide.pdf (accessed 1 May 2020).

Lourandos, H. (1983). Intensification: A Late Pleistocene-Holocene Archaeological Sequence from Southwestern Victoria. *Archaeology in Oceania, 18*(2), 81–94.

Massola, A. (1968). *Bunjil's Cave: Myths, legends and superstitions of the Aborigines of South-East Australia.* Melbourne: Lansdowne Press.

McBryde, I. (1978). *Wil-im-ee Moor-ing:* Or, Where do Axes Come from? *Man, 11,* 354–382.

Morieson, J. (1966). *The Night Sky of the Boorong: Partial Reconstruction of a Disappeared Culture in North-West Victoria.* Unpublished MA Thesis, The University of Melbourne, Melbourne.

Neidjie, B., Davis, S., & Fox, A. (1985). *Kakadu man ... Bill Neidjie.* Queanbeyan: Mybrood Books.

O'Brien, L. Y. (2007). *And the Clock Struck Thirteen: The Life and Thoughts of Kaurna Elder Uncle Lewis Yerloburka O'Brien.* Kent Town: Wakefield Press.

Paradies, Y. (2019). *Embedding Indigenous Engagement in Deakin Courses.* Unpublished Paper Presented at the Deakin University Academic Board.

Powell, B., Tournier, D., Jones, D. S., & Roös, P. B. (2019). Welcome to *Wadawurrung* Country. In D. S. Jones & P. B. Roös (Eds.), *Geelong's Changing Landscape: Ecology, Development and Conservation* (pp. 44–84). Melbourne: CSIRO Publishing.

Rose, D. B. (1988). Exploring an Aboriginal Land Ethic. *Meanjin, 47*(3), 378–387.

Rose, D. B. (1992). *Dingo Makes us Human: Life and Land in an Aboriginal Australian Culture*. Cambridge: Cambridge University Press.

Rose, D. B. (1996). *Nourishing Terrains: Australian Aboriginal Views of Landscape and Wilderness*. Canberra: Australian Heritage Commission.

Rose, M. (2017). The Black Academy: A Renaissance Seen Through a Paradigmatic Prism. In I. Ling & P. Ling (Eds.), *Methods and Paradigms in Education Research* (pp. 326–343). Philadelphia: IGI Global.

Sanyal, S. (2012). *Land of the Seven Rivers: A Brief History of India's Geography*. Gurgaon: Penguin.

Tawa, M. (1996, March). Liru and Kuniya. *Architecture Australia*. Available at: https://architectureau.com/articles/liru-and-kuniya/. Accessed 1 June 2019.

Thadoor, S. (2015, July 24). Read: Shashi Tharoor's Full Speech Asking UK to Pay India for 200 Years of Its Colonial Rule. *News 18*. Available at: https://www.news18.com/news/india/read-shashi-tharoors-full-speech-asking-uk-to-pay-india-for-200-years-of-its-colonial-rule-1024821.html. Accessed 1 May 2020.

Thadoor, S. (2017). *Inglorious Empire: What the British Did to India*. Melbourne: Scribe Publishing.

Tunbridge, D. (1988). *Flinders Ranges Dreaming*. Canberra: Aboriginal Studies Press.

Victoria. (1974). *Deakin University Act 1974* (Vic).

Victoria. (2006). *Aboriginal Heritage Act 2006* (Vic).

Victoria. (2009). *Deakin University Act 2009* (Vic).

Zola, N., & Gott, B. (1992). *Koorie Plants Koorie People: Traditional Aboriginal Food, Fibre and Healing Plants of Victoria*. Melbourne: The Koorie Heritage Trust.

CHAPTER 6

Student and Graduate Voices

David S. Jones, Kate Alder, Shivani Bhatnagar,
Christine Cooke, Jennifer Dearnaley, Marcelo Diaz,
Hitomi Iida, Anjali Madhavan Nair, Shay-lish McMahon,
Mandy Nicholson, Gavin Pocock, Uncle Bryon Powell,
Gareth Powell, Sayali G. Rahurkar, Susan Ryan,
Nitika Sharma, Yang Su, Saurabh V. Wagh,
and Oshadi L. Yapa Appuhamillage

D. S. Jones (✉) • U. B. Powell
Wadawurrung Traditional Owners Aboriginal Corporation,
Geelong, VIC, Australia
e-mail: davidsjones2020@gmail.com

K. Alder
Maribyrnong City Council, Melbourne, VIC, Australia

S. Bhatnagar • S. V. Wagh
Moir Landscape Architecture, Islington, NSW, Australia

C. Cooke • S. Ryan
School of Architecture & Built Environment, Deakin University,
Geelong, VIC, Australia
e-mail: ccooke@deakin.edu.au; rsusa@deakin.edu.au

© The Author(s), under exclusive license to Springer Nature
Singapore Pte Ltd. 2021
D. S. Jones (ed.), *Learning Country in Landscape Architecture*,
https://doi.org/10.1007/978-981-15-8876-1_6

To address this inadequacy of longevity, reflective and contextual data, this research project invited 20 postgraduate students and graduates near and after the completion of their MLArch or MPlan course four qualitative judgment questions. Participant selection was random by availability and willingness, so opt-in participant involvement. The survey occurred over

J. Dearnaley
Balyang Consulting, Geelong, VIC, Australia

M. Diaz
MINT Pool and Landscape Design, Melbourne, VIC, Australia
e-mail: marcediaz9940@gmail.com

H. Iida
Kihara Landscapes, Melbourne, VIC, Australia

A. M. Nair
Ground Ink, Sydney, NSW, Australia

S.-l. McMahon
GHD Woodhead, Melbourne, VIC, Australia
e-mail: Shay.McMahon@ghd.com

M. Nicholson
Tharangalk Art, Melbourne, VIC, Australia

G. Pocock
Garden Consultants, Geelong, VIC, Australia

G. Powell
Wadawurrung Traditional Owners Aboriginal Corporation,
Geelong, VIC, Australia

Legals Lawyers and Barristers (LLB) Pty Ltd, Canberra, ACT, Australia

S. G. Rahurkar
Inspiring Place, Hobart, TAS, Australia
e-mail: sayalirahurkar@gmail.com

N. Sharma
Mexted Rimmer Landscape Architecture, Geelong, VIC, Australia

Y. Su
Landscape Architect, Melbourne, VIC, Australia

O. L. Yapa Appuhamillage
Thomson Hay Landscape Architecture, Ballarat East, VIC, Australia

January–March 2019, of which all participants were in various stages of their studies or completion and all had successfully completed this unit upon their first attempt. Some 13 detailed responses were forthcoming and their authors are from a breadth of cultural backgrounds (Australian, Sri Lankan, Indian, Japanese, Chinese) and one Aboriginal from *Eora* and *Awabakal Country* descent, with a majority being female, all of which statistically reflects the domestic/international cultural profile of this course. There is also one online participant in the respondents. Two students had completed their MLArch and MPlan studies and had selected postgraduate research studies, and you can discern the role the unit may have had on their postgraduate research topic selection in their clear statements.

The data in the following four tables (Tables 6.1, 6.2, 6.3 and 6.4) has not been edited, grammatically amended or spellchecked, and the outcomes are progressively discussed. The data is sequential: de-identified participant codes listed upon receipt enabling thought comparison between tables and any identifier titles or project names have been deleted to ensure anonymity.

To the question 'What was the most important ideas and or information that you learnt in the SRL733 unit?', as set out in Table 6.1, all participants were extremely positive in both themselves and their practice outlook to the world, or their new 'global citizenship' ethical position.

In these responses, there was not a note that the content they had learnt in the unit had no value to their career, and thus the content was making a powerful role in their personal learning journey. There was considerable self-reflecting in their answers. They talked about their evident shift of values or perspectives arising from the unit, not linked changes to any other unit in the course. There were also aspects of personal guilt, desire to be respectful, a desire to make positive respectful changes now with these new values and perspectives. While references to Aboriginal knowledge and *Country* were evident, there was also clear reference to a raft of Indigenous peoples that had either some past experiences with or were present in their home nation.

To the question 'Has/Did the unit SRL733 influence/d your choice of Masterclass and or Thesis topic(s)?', as set out in Table 6.2, there were positive responses in personal thought, but mixed responses in terms of use and implementation.

To explain, the mixture of responses, in this course postgraduate students do two capstone projects, that do not necessarily have to be linked, of which the Masterclass project is fixed in Trimester 2 as a culmination design capstone. Masterclass comprises 50% of a student's workload in that Trimester, is strategically the design-based capstone unit where the

Table 6.1 Q1: What was the most important ideas and or information that you learnt in the SRL733 unit?

#	Respondent cultural background	Response
SRR1	Indian	… the outlook that we/I as human being and as a designer should have towards people of indigenous communities all over the world. Additionally, I realised the importance the role a community places in development of a place. I understood the importance of community, environment and country for an individual of indigenous community. I realised that designers should have a conscious understanding of a community and culture while designing a project such as memorial, parks etc.
SRR2	Australian	Changed the perception that I grew up with about the Australian landscape being a harsh and unforgiving place that has to be worked hard (i.e. flogged and interfered with massively) to be productive. I now appreciate that indigenous communities were able to live in sync with the land and to keep it productive for at least 40,000 years, in contrast to "Western" ways of exploiting the land that have seen massive degradation of the environment (soils, waters, air, flora and fauna diversity, etc.) in just over 200 years.
SRR3	Sri Lankan	We all are followers of our own indigenous people. Unconsciously we are doing the same thing in a different way what they have done. Adaptability and capability are the two facts I would like to raise among the whole bunch of aboriginal information. Those people were incredibly adaptable to a harsh atmosphere. At the same time, they were capable to develop a unique code to express their culture.
SRR4	Indian	The SRL733 unit as a foundation unit was extremely beneficial in setting up a strong theoretical base that helped many students understand the relevance of Indigenous cultures and to consider factors related to them in the very early stages of landscape design/planning. As an integral aspect of the Australian landscape, it is critical for students of landscape architecture / planning to engage with the meanings that Indigenous cultures find in their country since they are the true custodians of their lands and are its rightful owners. … The unit did not limit itself to the Australian context and provided enough flexibility for students to study Indigenous cultures in international contexts as well, and this was a good way for students to understand that the approach/protocols to consult and involve them in planning/design decisions is not limited to any particular context.

(*continued*)

Table 6.1 (continued)

#	Respondent cultural background	Response
SRR5	Australian	Although 6 years have transpired since undertaking this unit, I am still astonished at the profound knowledge that was transferred throughout this unit and dismayed that my early education either at primary or secondary level didn't even consider the Indigenous reality of their history in Australia. My memories of history classes are glorified colonisation stories and those of the early settlers. But to discover how ancient and managed the Australian landscape has been for tens of millennia in a sustainable and sophisticated approach is not only startling but compelling and should be regarded as an exemplar of best practice management techniques of this ancient Australian landscape. This unit went beyond introducing concepts of Indigenous culture including notions of country, season, time, language and myth. It imbued a significant mindset shift that inspired a deeper meaning in this country
SRR6	Japanese	Aboriginal people see their selves as custodians of the land. This is the most important idea for us grown up urbanized areas. It seems that cities have been developed with more egoistic mind compared to indigenous people's ways. Now we are going back to their 'sustainable' way of using the land. Therefore, that idea is the most valuable and meaningful for landscape students.
SRR7	Aboriginal (*Awakabal*)	This unit further cemented pre-existing beliefs and knowledges. As a past student of this unit it really helped me articulate knowledge into the written form—As an Aboriginal person this unit further emphasised that we as Aboriginal people identify ourselves with various elements i.e., land sea sky and are all relational components of Country. This unit looked at these ideas through a multi-dimensional lens that was inclusive of all elements that embracing origins but considered contemporary views … I think the most important information that was taught in this unit is that the idea of the importance of process and involvement and also that oral history and communication are a way for us to understand the past, present and future

(*continued*)

Table 6.1   (continued)

#	Respondent cultural background	Response
SRR8	Indian	... I was exposed to the world of Aboriginal story telling. These verbal stories helped understand the natural history of the land and how the land unfolded through time. Some of the key aspects of how the Aboriginal people used their collective knowledge of about 60,000 years to manage the land (example, fire link to management and fire linked to human health), healing the land and threading the land lightly. The importance of the deep connection of the traditional owners of the land with country. This reflects the idea that one's wellbeing is intertwined with the wellbeing of their country.
SRR9	Australian	(a) Understanding Indigenous peoples and their cultural and spiritual relationships to land (Country), language, name, knowledge transferal. (b) Understanding and developing planning and management tools and approaches to understanding of Indigenous protocols
SRR10	Indian	The Indigenous population of Australia considers the land as their country. They talk about country as one would talk about a person. It includes the earth, sea, sky and everything within; not only in terms of tangible but also the intangible aspect of country. The people have a responsibility to their country as the land is not just their future, but past and present as well ... another important teaching from this unit is the importance of the understanding of a native language to understand a place and for its further development. Unlike western place names, native place names have the potential to recite the nature of landscape present in the place along with the type of resources and the possible cultural and spiritual connection to the people. This information can help designers to design spaces considering the potential opportunity or concern presented by the place.
SRR11	Australian	An appreciation of the sophistication of the indigenous communities' ability to read 'The land' and utilise land's resources in harmony with the landscape. An appreciation of the destructive aspect of Western notions of environmental land management, and how much we can learn from the original inhabitants.

(continued)

Table 6.1 (continued)

#	Respondent cultural background	Response
SRR12	Indian	… I learnt about the connection of people to a place or what a 'country' means to indigenous people. I learnt to peruse intricate relationships between the environment and its people and discern responses that mold our communities and places. Based on my understanding of this unit I bolstered a sense of integrity and developed a rectitude towards people and their connections to landscapes.
SRR13	Chinese	The way how indigenous people interact with their own nation, the nature and culture, as well as the spirit of valuing and smartly exploring nature, the respect of culture and its inheritance, in many ways influence me in the landscape design process both theoretically and practically.

last two years of studies culminates in a student self-selected and driven project, with team tutor supervision, and is undertaken and presented to a joint academic-practice jury. In contrast, the thesis is a self-directed research project about a topic of student choice. This unit comprised 50% of a student's workload in that trimester, was curated with one tutor supervisor, that explores and interrogates a topic, resulting in a paper of up to 10,000 words. Where an Aboriginal-related topic is/was selected, the research project is/was human ethics review by the upper-level Deakin University Human Research Ethics Committee given that all 'Aboriginal' topics are deemed 'greater than low risk' in their scope and research risk.

A key feature of academic research standing is its 'contribution to knowledge' in a realm. Coursework and higher degree Theses, Masterclass Projects and associated publications are included in this 'body' as they individually and/or collectively contribute to a 'body of knowledge' on a discipline and in this case they are generally geographically relevant to *Wadawurrung Country*.

In this instance, the research inspired by SRL733 since has led to research on the following topics either directly relevant to *Wadawurrung Country*, to another *Country*, or to a theme pertinent to IKS. These include the subtopics of: environmental systems (Eccles and Jones 2020; Roös 2015, 2017, 2020; Powell et al. 2019; Pocock 2013; Pocock and Jones 2013; Wadawurrung 2019); terrestrial and aquatic ethnobotany

Table 6.2 Q2: Has/Did the unit SRL733 influenced your choice of Masterclass and or Thesis topic(s)?

#	Respondent cultural background	Response
SRR1	Indian	Initially, this unit influenced my choice of topic for thesis but I wasn't able to come with a concrete idea for my thesis. Hence, I decided to change the topic.
SRR2	Australian	I had already decided my thesis topic before doing the unit. It will influence my Masterclass topic, I think, or at least my treatment of it.
SRR3	Sri Lankan	My topics did not have direct influence, although the research had inspired by aboriginal knowledge of fire prevention to mine rehabilitation purposes. The masterclass project helped to conclude the aboriginal landscape characters (by emotions, enclosures, textures and forms) to the psychological healing of drug-addicted patients.
SRR4	Indian	To a large extent, yes. Since I developed a keen interest in the unit, I tried to apply the knowledge that I gained in SRL733 in subsequent units. My Masterclass and Thesis topic delved in ecological land management and planning techniques at the regional scale. The site that I chose carries an interesting and vibrant Aboriginal as well as colonial past, and consideration of these factors for research and design was important. It was not the central theme/topic of discussion for my project, but the information that I gathered with regards to Aboriginal land management practices played a very important role in identifying the layered meanings hidden in the landscape, which inherently contributed towards the outcome of my Thesis report and Masterclass design intervention.
SRR5	Australian	Yes, both my MasterClass and Thesis topics involved designing and restoring Indigenous gardens and parklands. MasterClass—…which is about remembering and respecting the local indigenous history, and to re-appropriate this site for a modern culture that's steeped in a forgotten culture. Thesis—…essentially I designed a physic garden that can educate modern society in how the Wadawurrung people used the indigenous plants for healing. Undertaking the research for this project I identified over 150 uses of plants that were used by the Wadawurrung people for medicinal purposes.

(*continued*)

Table 6.2 (continued)

#	Respondent cultural background	Response
SRR6	Japanese	Yes. It has influenced my design process for the projects since I had this unit. Before having this unit, it was too hard to concern indigenous culture aspects in my design although I tried to show respect to them. What I could do is just include Aboriginal art in my design, but now my design can have indigenous narrative and it enriches whole design intent
SRR7	Aboriginal (*Awakabal*)	It didn't influence the topic of choice; however, it did influence the way in which it was approached and the chosen methodology.
SRR8	Indian	The deep connection between the traditional owners of the land with country helped them manage their nation in a balanced sustainable way for 60,000 years. Issues such as pollution and climate change is the result of the disconnection between the local communities and their country. Many researches had looked into the scientific findings of how to restore landscapes, rivers and clean pollution. But for this restoration to sustain the local communities must relink both the people's experience and spiritual connection to the landscape itself. In ancient India people were spiritually and culturally connected to the rivers and their country. I was able observe these common traits of narratives, beliefs and knowledge systems between the Indigenous communities of Australia and people of ancient India. Through the masterclass project ... I had tried to implement the lessons I learnt from SRL 733. Though it might not be possible to bring back the landscape to its once glorious state. The Indigenous communities have proved that the healing of landscape is possible through cultural activities and connection with country. ... narratives is a big part in the design process, and it was reflected through architecture. From the start of the journey along the river to the peak of the journey, people would be exposed to different experiences. These experiences were carefully designed to reflect the Indian mandala concept to bring unity within the man and the ecosystem (where man is just one element among the whole ecosystem).
SRR9	Australian	Undoubtably, as gaining an understanding of Indigenous people and their culture opened my eyes to issues within our Victorian planning scheme and cases around cultural landscape, specifically spiritual relationships to land. SRR732 research methodologies allowed me to 'test the water' with a case study around cultural landscapes, which lead to my thesis ...

(continued)

Table 6.2 (continued)

#	Respondent cultural background	Response
SRR10	Indian	Unit SRL733 have opened my mind to a new world of Indigenous belief and knowledge. Throughout Australia, even in the arid deserts of Australia, the Indigenous population have developed an intensive knowledge of food and water sources embedded in their knowledge of seasons and seasonal variations. Across Australia, the aboriginal communities have sophisticated yet sustainable land and water management systems which helped them to survive during climate change resulting in hostile climatic conditions.
SRR11	Australian	Sadly no. Well not directly. There is much in common between Japanese and Aboriginal concepts of God's (Japanese concept of kami's) and respect for nature and the land. This topic really should be explored one day post Ph.D.
SRR12	Indian	… had a strong influence on all my assignments for at Deakin University in subsequent trimesters. This unit induced a resolute philosophical design approach towards selecting a topic for masterclass, where the aim/objective was to create awareness and respect the existing Indigenous people belonging to countries on either side of …, and further inferring integrated outcomes through my research thesis where I drew initiatives from indigenous theories of fire ecology to manage …'s wetland systems
SRR13	Chinese	Yes, it inspires me in designing aboriginal communities in the masterclass project. It's not just limited to aboriginal community design, but it also broadens my perspectives of design in bridging connections between nature and culture, exploration of hidden spirits of being modest and respectful of nature and etiquette in the design process.

(Dearnaley 2015a, b, 2019; Swan 2017; Thurstan et al. 2018); anthropological patterns and processes (Threadgold 2020); places and place-making (Allwood 2018a, b; Ryan 2016; Su 2014, 2016); *Country* (Nicholson 2020; Nicholson and Jones 2020; Powell et al. 2019; Powell and Jones 2018; Wadawurrung 2019); designing narratives (Bhatnagar 2017, 2018; McMahon 2017; Nicholson et al. 2020, 2018; Rahurkar 2019; Tang 2013, 2014; Wagh 2017, 2018; Whelan 2014; Yapa Appuhamillage 2019), perspectives and values (Brown 2018; Cooke 2018; Powell et al.

Table 6.3 Q3: Has this unit SRL733 changed your perspective about Indigenous communities, their values, and their Knowledge Systems? And how?

#	Respondent cultural background	Response
SRR1	Indian	This unit helped me to understand the level of respect the people of indigenous community deserve from everyone. While learning about their past predicaments, I developed respect for them which indecently helped me to broaden my outlook as a designer. I understood the importance the elders of community play in spreading the knowledge that they have gained through experience.
SRR2	Australian	It has definitely heightened my appreciation of the importance of their oral tradition and marks on the land in effectively passing on a huge and vital body of knowledge over many generations. This goes for indigenous people all around the world who have been very well adapted to their environment through the passing on of oral knowledge. It has also made me more aware of Australian indigenous groups' continuing and deep connection to the land and its spirits. It has also made me realise that a system of knowledge does not need to stem from orthodox science and the scientific method etc. to make ecological sense. Indigenous people had their own system for doing science, and it worked. Finally, it has really underlined what a shabby con act the whole terra nullius story was. It was just a convenient pack of lies on the part of the colonial office, successive waves of settlers and Australia's governmental/legal infrastructure.
SRR3	Sri Lankan	Indigenous people all over the world had minimum resources and technology to survive. They have achieved with the sense that is missing to us today. They knew about the rain, clouds, soil, and aliveness. They had fear of nature as well their protector also nature. Native customs, beliefs, and values develop around the natural elements or spiritual spaces. They had their rules to protect it and continue it as we do now. It was extremely amazing to meet some indigenous people in the classroom. They are not different, but they think differently about the landscapes.
SRR4	Indian	Yes, SRL733 helped me gain a better perspective on indigenous communities and helped me identify the reason why they should be approached, consulted and involved in a manner that is sensitive towards their cultural knowledge, values and the meanings they find in their country.

(*continued*)

Table 6.3 (continued)

#	Respondent cultural background	Response
SRR5	Australian	Upon completion of my Masters in Landscape Architecture, my path led into a PhD that explored Indigenous Ecological Knowledge at a deeper level. My perspective has changed from being initially ill informed and ignorant of the knowledge systems and values, protocols and cultural diversity of the Australian Indigenous people to having a somewhat deeper level of understanding that enabled me to write a PhD thesis.
SRR6	Japanese	This unit provided us with a lot of opportunities to contact with Aboriginal materials such as possum skin clothes, hunting tools and so on. Simply it was very enjoyable, and it is the best way of start learning culture! I wish there were this kind of chances to be associated with aboriginal culture opened up for tourists / backpackers as I was the one.
SRR7	Aboriginal (*Awakabal*)	This unit changed my perspective of Aboriginal communities outside of my own. As this unit was heavily influenced by the local Aboriginal community here in Geelong it allowed me to gain a better understanding of where I came too—after relocating from another state and a completely different community—This unit allowed me to place this community into a way that I could understand better
SRR8	Indian	Through SRL733 I became aware of the diversity in values and knowledge systems (example: Seasonal calendars) in Indigenous communities from different nation groups. This diversity brought forward the collective knowledge of different aboriginal nations in Australia. This changed my perspective of Indigenous people from communities of nomads who lived only for survival to Indigenous communities with their own evolved culture and rich history. Aboriginal ancestral narratives aren't just about the land, but it is also about the sky. They used the changing patterns of the stars to predict seasonal changes, availability of certain foods. These knowledge systems were also used to navigate through space. Understanding the Aboriginal Knowledge system provides vital information about the underground water sources, climate change of the landscape and treading the earth lightly.

(continued)

Table 6.3 (continued)

#	Respondent cultural background	Response
SRR9	Australian	Though entering SRL 733 with a broad understanding and interest in our country's Indigenous history, my background is of British heritage and a Year 12 Australian History student in the early 1990s which both emphasised our 'terra nullis' history. SRL 733 altered my perspective, enhanced my knowledge and further developed my respect for Indigenous cultural and spiritual connection to land.
SRR10	Indian	I was first exposed to Lake Tyrrell, and its importance in terms of Indigenous cosmology and dreaming stories, during the unit SRL733. … I developed an interest in the lake and the fact that it is one the few places where you can see the entire milky way galaxy due to the absence of light pollution. … I went on to develop the project with the aim of preserving and enhancing the environmental assets, improving tourist experience and visitors' awareness about the region and its aboriginal heritage by providing a series of lookouts along the Lake Tyrrell at strategic locations to promote the visual experience of the Salt Lake in the day and stargazing in the night time.
SRR11	Australian	YES INDEED. By doing field trips and having guest lecturers it was a creative and rich educational experience. I got to travel to central Australia and learn about MURIEL Pink and those that saw the great value of aboriginal culture.
SRR12	Indian	SRL 733 has taught me to be appreciative and comprehend a complex relationship between humans and nature. The unit helped me acknowledge that 'landscapes' extend beyond a mere horizontal plane. It deepened my understanding of inter-relations between places/landscape and the effect of curated human responsibilities on such landscapes. It has led me to appreciate the connection of people to land and enrich my approach to planning and environmental designs. It has made me aware and respect all Indigenous communities (traditions, knowledge etc.). This has aided in a philosophical approach created through respect, integrity and diligence towards indigenous cultures.

(*continued*)

Table 6.3 (continued)

#	Respondent cultural background	Response
SRR13	Chinese	This unit instructs my deeper understanding of the dependable but innovative connections between indigenous communities and nature, their incredible knowledge systems enlighten me in appreciating the nature and landscape, and valuing how culture and spirits immersed in their daily life. Those strong connections between human and nature, respect of land, inheritance of ceremonies and culture encourage me creating new insights of landscape design in the future.

2017; Rahurkar 2020; Ryan 2017; Stallard 2011), and archaeoastronomy and place (Bhatnagar 2016; Sharma 2018) (Figs. 6.1, 6.2 and 6.3).

We are also conscious that many insightful Australian-based texts have been released over the last 10 years that are directly relevant to teachings in this class. These include Australia (2017), Cahir et al. (2018), Gammage (2011), Gunditjmara People with Wettenhall (2010), MAC (2016), Neale (2017), Parks Victoria (2015), Pascoe (2014), Pieris et al. (2014), Powell et al. (2019), Steffensen (2020), Watson (2019) and YRNTBC (2011).

Collectively, the participant's responses indicated a strong willingness to engage with or somehow incorporate engagement and Indigenous knowledge in either the Thesis or Masterclass project, but that their personal choice of topic or location may have negated the opportunity to entertain this.

To the question, 'Has this unit SRL733 changed your perspective about Indigenous communities, their values, and their Knowledge Systems? And how?' as set out in Table 6.3, all participants stated that the unit had a major learning role in changing their values and perceptions to Indigenous issues and Indigenous Knowledge Systems.

Statements like, 'helped me to understand the level of respect' and 'has definitely heightened my appreciation' and 'It has led me to appreciate the connection of people to land and enrich my approach to planning and environmental designs' culminated in an insightful comment by one of the Sri Lankan students that 'They are not different, but they think differently about the landscapes'. Not one respondent stated that the content they

Table 6.4   Q4: Thinking as a graduate / practitioner, how important is the SRL733 learning content for your work activities?

#	Respondent cultural background	Response
SRR1	Indian	As a graduate, I have not had opportunities to delve into the importance of this unit in context of work activities but I believe that the insight I developed during the course of studying this unit would be extremely helpful in future. While working, I will keep in mind the importance of connection to country and indigenous environment for the first people of this nation. After studying this unit, I aspire to be a sensitive and conscious designer in future.
SRR2	Australian	Very important, in giving me an appreciation of the depth of indigenous cultures and knowledge before engaging with groups on projects. It has, though, made me a bit more wary of engaging with traditional owners without firstly being properly prepared and educated. It is so important to appreciate their local connections to country, and not to assume that all groups had the same cultural background and traditions. On the other hand the introduction to procedures such as the smoking ceremony and the importance of greetings, welcomes and exchanges has given me a better insight into building rapport with traditional owners.
SRR3	Sri Lankan	First sight, oh it is the history of people in Australia. Then no it is the base of Australia. This unit gave the sense of how people move with history unconsciously and who are the few who follow the sense of history. They have used sounds of nature to communicate, they have used forms of nature to communicate, and they made own languages, territories, customs for sake of their own clans the same as every indigenous clan. Designers inspiring from the past for the future. Because of past unconsciously familiar to everyone and understandable.
SRR4	Indian	Acknowledgement, assessment, evaluation and application of the knowledge obtained from Indigenous cultures is critical. For example, large scale master planning interventions can hamper existing ecological processes which will in turn affect the processes of land parcels in the surrounds and those that extend beyond. These are the primary reasons for constantly evolving ecological behaviour, which is essentially the way nature behaves. Understanding the land's inherent processes and how to tackle them can help gain insight on future directions for its management, and that's the knowledge that Indigenous cultures receive from their respective ancestors through Dreamtime stories, cultural practices and layered meanings. ... In many ways this unit has aspired me to carry this knowledge and try to propel a change in the way practitioners approach their projects

(*continued*)

Table 6.4 (continued)

#	Respondent cultural background	Response
SRR5	Australian	The content from this unit has irrevocably altered the direction of my life. I was planning to practice Landscape Architecture which is why I subscribed to this Master's programme, however I am compelled to continue my research in the area of Indigenous Ecological Knowledge and learn how to care for Country and how best to support and restore this land we call Australia.
SRR6	Japanese	Assignments made me go back to my culture, it was great opportunity to learn and introduce it to the people. The learning content of this unit has certainly enriched my design process and ethical thinking as a landscape architect.
SRR7	Aboriginal	The content of this unit identifies that as professional planners we should and should continue to gain knowledge and new knowledges so that no wrong doings are repeated again. Further to that, it is the process of involvement is more important then the end results.
SRR8	Indian	As a landscape architect is it important understand of the history of landscape, flora and fauna and the belief systems of the local communities. SRL 733 provided us the competence to engage Indigenous communities in design and planning processes. This would enable us graduate/ practitioners to provide solutions with the understanding of the two worlds (indigenous communities and the present urban setting).
SRR9	Australian	To achieve wholistic outcomes, with cultural richness and respect, engagement with Indigenous communities in design, planning and protocol is critical. However, as a planning graduate and practitioner, SRL 733 and the knowledge it delivers is not compulsory in our studies or profession. Unfortunately, this continues our focus, both as professionals and a society, centred around colonial systems without due regard to those who were before it
SRR10	Indian	The Indigenous population, throughout the world, believe in the responsibility to their land where they look after the health and the vitality of the land as it provides them resources for their sustenance. As future practitioners, it is our responsibility to take care of the land. Present day practitioners can develop a more efficient development methods by tapping into the immense knowledge of environment and land management methods used by most of the Aboriginal communities around the world. However, care must be taken to maintain the localised essence of indigenous knowledge as no two places or communities are the same. ... the biggest lesson that I have learnt in this unit is that, we could either start fresh or use their thousands of years' worth of knowledge as a starting point while moving forward and developing our cities and environment.

(continued)

Table 6.4 (continued)

#	Respondent cultural background	Response
SRR11	Australian	Similarly answered above, it added richly to my overall MLA education and to gain an appreciation of other countries Indigenous people and understanding of their land management including the Ainu 蝦夷 of Japan.
SRR12	Indian	As a graduate and a professional, this units' learning contents have defined my design approach. An integrated design approach philosophy has been orchestrated through my ability to pursue indigenous relevant information to instruct built environment principles.
SRR13	Chinese	As a practitioner, it is important to learn from nature, creatively utilising indigenous unique traditional knowledge in sustainably exploring nature, respect of their rich spirits and diverse culture hidden in the landscape through histories, and value of their strong connections within clans, etc. all of these criteria bring insight in landscape design processes.

Fig. 6.1 Cross-section in Restoration of Yollinko Park, designed by Jennifer Dearnaley, Highton, Vic. (Image author: Jennifer Dearnaley 2015a, b)

Fig. 6.2 Masterplan design for proposed Wathaurong Cultural Centre, Anakie, Vic. (Image author: Brian Tang 2013)

had learnt in the unit had no value to their career, and thus the content had had a powerful role in their personal learning journey.

The other point to note in the responses is the knowledge qualities that each respondent often cited were different to each other, like stars, Elders, plants, colonised histories, seasonal calendars. This demonstrated that the current breadth of the unit's teaching content was also capturing diversity of interests and inspiration of student and interests. This is important, given that topics cited interestingly did not coalesce around 2–4 classic Australian architectural examples, like the Brambuk Cultural Centre at *Gariwerd* (The Grampians) or the Uluṟu-Kata Tjuṯa Cultural Centre at *Uluṟu*. Both projects were co-authored by architect Greg Burgess. Most architecture students passionately knew these as national exemplars without knowing the real design narratives behind their structures, their landscapes and their contextual positionings.

Fig. 6.3 *Anakie Youang* / The Anakies [comprising *Coranguilook*, *Baccheriburt* and *Woollerbeen*], Anakie, Vic. (Image author: Susan Ryan 2017)

There was also a sense or tinge of anger in their respective secondary schooling, whereby they had been taught classic colonised histories of the Australian, Indian, Sri Lankan or Japanese respective landscapes, and that there was clearly decolonised versions that offered new design insights and inspirations for their possible future applications.

The last question, 'Thinking as a graduate / practitioner, how important is the SRL733 learning content for your work activities?', as set out in Table 6.4, charted a future projection in their viewpoints.

Several of the respondents had already picked up employment positions in offices or agencies or had entered into further research degrees. The question was consciously couched to project their thoughts into the future, recognising that normal 'student evaluation of learning and teaching' is centred in the now and the immediacy of when a survey is being completed. Again, there was marked commonality in responses. Thus, statements like 'I believe that the insight I developed during the course of studying this unit would be extremely helpful in future' and '... has irrevocably altered the direction of my life' and '... learning content of this unit has certainly enriched my design process and ethical thinking as a landscape architect'. One respondent philosophically stated, 'The biggest lesson that I have learnt in this unit is that, we could either start fresh or use their thousands of years' worth of knowledge as a starting point while moving forward and developing our cities and environment'. There was not a note that the content they had learnt in the unit had no value to their career, and thus the content in their eyes was creating a powerful role in their personal learning journey. There was a positive outlook in all these responses, and a desire to better serve as 'global citizens' that goes back the deeper intent of that 'global citizenship' CLO goal.

References

Allwood, B. (2018a). *Country, Memory and Remembrance. Reconciliation Through Intercultural Design in Landscape Architecture: A Case Study from Far North Queensland*. Unpublished SRR711 MLArch Thesis, Deakin University, Geelong.

Allwood, B. (2018b). *White Rock Cultural Domain*. Unpublished SRD768 Landscape Masterclass Project, Deakin University, Geelong.

Bhatnagar, S. (2016). *Narrating the Night: A Guide for an Astronomy Trail in Lake Tyrrell, Precinct, Victoria*. Unpublished SRL733 Assignment, Deakin University, Geelong.

Bhatnagar, S (2017), *Lifting the Veil: A Master Plan for Lake Conneware, Bellarine Peninsula*. Unpublished SRD768 Landscape Masterclass Project, Deakin University, Geelong.

Bhatnagar, S. (2018). *Re-imagining Management of Ecological Patches and Corridors: Envisioning the Future of Bellarine Peninsula, Victoria*. Unpublished SRR711 MLArch Thesis, Deakin University, Geelong.

Brown, P. (2018). *The Exploration of Connection to Country as a Potential Fourth Architectural Design Paradigm to Aboriginal Housing in Remote Australia*. Unpublished SRR711 MLArch Thesis, Deakin University, Geelong.

Cahir, F., Clark, I. D., & Clarke, P. A. (2018). *Aboriginal Biocultural Knowledge in South-Eastern Australia: Perspectives of Early Colonists*. Clayton South: CSIRO Publishing.

Cooke, C. (2018). Indigenous Fluency: In the Victorian Planning Profession. Unpublished MPlan Thesis, Deakin University, Geelong.

Dearnaley J. (2015a). *Restoration of Yollinko Park*. Unpublished SRD768 Landscape Masterclass Project, Deakin University, Geelong.

Dearnaley, J. (2015b). *Wathaurong Medicinal Plant Walk*. Unpublished SRR716 MLArch Thesis, Deakin University, Geelong.

Dearnaley, J. (2019). *Wadawurrung Ethnobotany as Synthesised from the Research of Louis Lane.* Unpublished PhD Thesis, Deakin University, Geelong.

Eccles, C., & Jones, D. S. (2020). Perspectives: Country, *Kuarka-dorla* (Anglesea) and Future (Eccles & Jones). In P. B. Roös (Ed.), *Heal the Scar – Regenerative Futures of Damaged Landscapes* (pp. 27–28). Geelong: The Live+Smart Research Lab, Deakin University.

Gammage, B. (2011). *The Biggest Estate on Earth: How Aborigines Made Australia.* Crows Nest: Allen & Unwin.

Gunditjmara People with G Wettenhall. (2010). *The People of Budj Bim: Engineers of Aquaculture, Builders of Stone House Settlements and Warriors Defending Country.* Heywood: em PRESS Publishing.

McMahon, S. (2017). *Post-Design Evaluation of Kitjarra Residences (Aboriginal Residences): Deakin University, Waurn Ponds.* Unpublished SRR711 MLArch Thesis, Deakin University, Geelong.

Murujuga Aboriginal Corporation (MAC). (2016). *Ngaayintharri Gumawarni Ngurrangga: We all come together on this Country – Murujuga Cultural Management Plan 2016.* Karratha: Murujuga Aboriginal Corporation.

Neale, M. (Ed.). (2017). *Songlines: Tracking the Seven Sisters.* Canberra: National Museum Australia.

Nicholson, M. (2020). *Being on Country off Country – Gunditjmara Country.* Unpublished PhD thesis, Deakin University, Geelong.

Nicholson, M., & Jones, D. S. (2018). Urban Aboriginal identity: "I can't see the durt (stars) in the city", in I. McShane, E. Taylor, L. Porter & I. Woodcock (Eds.), *Proceedings of Remaking Cities: 14th Australasian Urban History Planning History Conference 2018, RMIT University, Melbourne, 31 January – 2 February 2018,* (pp. 378–387). Available at: https://www.remakingcities-uhph2018.com/ + https://cloudstor.aarnet.edu.au/plus/s/g0FtJzRx3H5vSTb#pdfviewer (accessed 1 January 2019).

Nicholson, M., & Jones, D. S. (2020). *Wurundjeri-al Biik-u (Wurundjeri Country), Mag-golee* (Place), *Murrup* (Spirit) and *Ker-up-non* (People): Aboriginal Living Heritage in Australia's Urban Landscapes in KD Silva. In *Routledge Handbook on Historic Urban Landscapes of the Asia-Pacific* (pp. 508–525). Abingdon: Routledge.

Nicholson, M., Romanis, G., Paton, I., Jones, D. S., Gerritsen, K., & Powell, G. (2020). 'Unnamed as Yet': Putting *Wadawurrung* Meaning into the North Gardens Landscape of Ballarat. *UNESCO Observatory E-Journal Multi-Disciplinary Research in the Arts,* 6(1): vii–viii, 1–19. https://www.unescoejournal.com/volume-6-issue-1/ and https://www.unescoejournal.com/wp-content/uploads/2020/06/JONES-NICHOLSON-POWELL-ROMANIS-GERRITSEN-PATON-1.pdf. Accessed 15 June 2020.

Parks Victoria, Gunditj Mirring Traditional Owners Aboriginal Corporation (GMTOAC) and Winda Mara Aboriginal Corporation (WMAC). (2015).

Ngootyoong Gunditj Ngootyoong Mara South West Management Plan. Melbourne: PV. http://parkweb.vic.gov.au/explore/parks/mount-eccles-national-park/plans-and-projects/ngootyoong-gunditj-ngootoong-mara. Accessed at 1 June 2020.

Pascoe, B. (2014). *Dark Emu, Black Seeds: Agriculture or Accident?* Broome: Magabala Books.

Pieris, A., Tootell, N., Johnson, F., McGaw, J., & Berg, R. (2014). *Indigenous Place: Contemporary Buildings, Landmarks and Places of Significance in South East Australia and beyond.* Carlton: Melbourne School of Design, University of Melbourne.

Pocock, G. (2013). Redeeming Fire: The use of fire as a design tool in the Australian landscape, in ed. R Davies & D Menzies, *Shared Wisdom in an Age of Change 2013 Aotearoa New Zealand, IFLA50, Proceedings of the 50ᵗʰ IFLA [International Federation of Landscape Architects] World Congress, Sky City Convention Centre, Auckland, New Zealand, 10th–12th April 2013*, pp. 91–103. Wellington: NZILA. ISBN 978-0-473-24360-9. http://www.ifla2013.com/. Accessed at 1 June 2019.

Pocock, G., & Jones, D. S. (2013). Indigenous Landscape Change and Climate Change: The Historical Transformation of the Port Phillip Bay from an Indigenous and Landscape Architectural Perspective. In N Gurran, P Phibbs & S Thompson (Eds.), *Proceedings of the 10th International Urban Planning and Environment Association Symposium, (UPE10 – NEXT City: Planning for a New Energy & Climate Future), Sydney, 24–27 July 2012* (pp. 129–147). Sydney: ICMS Pty Ltd. http://www.upe10.org/. Accessed at 1 June 2019.

Powell, G., & Jones, D. S. (2018). *Kim-barne Wadawurrung Tabayl*: You Are in *Wadawurrung Country. Kerb: Journal of Landscape Architecture, 26*, 22–25.

Powell, G., McMahon, S., & Jones, D. S. (2017). Aboriginal Voices and Inclusivity in Australian Land Use *Country* Planning. In P. K. Collins & I. Gibson (Eds.), *Proceedings of the International Conference on Design and Technology 5–8 December 2016, Deakin University, Geelong* (pp. 30–36)., at http://knepub-lishing.com/index.php/KnE-Engineering/issue/view/43. Accessed at 1 June 2019.

Powell, B., Tournier, D., Jones, D. S., & Roös, P. B. (2019). Welcome to *Wadawurrung* Country. In D. S. Jones & P. B. Roös (Eds.), *Geelong's Changing Landscape: Ecology, Development and Conservation* (pp. 44–84). Melbourne: CSIRO Publishing.

Rahurkar, S. G.. (2019). *Trailing Through Land and Water: Great Ocean Road Aboriginal Interpretive Trails.* Unpublished SRD768 Landscape Masterclass Project, Deakin University, Geelong.

Rahurkar, S. G. (2020). *Empowering Indigenous Stories in the Landscapes of Melbourne and Geelong.* Unpublished SRR711 MLArch Thesis, Deakin University, Geelong.

Roös, P. B. (2015). Indigenous Knowledge and Climate Change: Settlement Patterns of the Past to Adaptation of the Future. *International Journal of Climate Change: Impacts & Responses, 7*(1), 13–31.

Roös, P. B. (2017). *Regenerative-Adaptive Design for Coastal Settlements: A Pattern Language Approach to Future Resilience.* Unpublished PhD Thesis, School of Architecture & Built Environment, Deakin University, Geelong. www.dro.deakin.edu.au/eserv/DU:30103436/roos-regenerativeada ptive-2017.pdf. Accessed at 1 June 2019.

Roös, P. B. (2020). *Regenerative-Adaptive Design for Sustainable Development: A Pattern Language Approach.* Cham: Springer Nature.

Ryan, S. (2016). *Creative Places in South West Victoria.* Unpublished SRL733 Assignment, Deakin University, Geelong.

Ryan, S. (2017). *Conserving Our Indigenous Landscapes to Acknowledge Our Country's History.* Unpublished SRR711 Thesis, School of Architecture & Built Environment, Deakin University, Geelong.

Sharma, N. (2018). *Perpetual Direl.* Unpublished SRD768 Landscape Masterclass Project, Deakin University, Geelong.

Stallard, K. E. (2011). *Where to Begin? An Exploration of How Architects Approach Aboriginal Housing in Australia.* Unpublished Master of Architecture Thesis, School of Architecture & Built Environment, Deakin University, Geelong.

Steffensen, V. (2020). *Fire Country: How Indigenous Fire Management Could Help Save Australia.* Richmond: Hardie Grant Travel.

Su, R. J. (2014). *Narana Creations.* Unpublished SRL733 Assignment, Deakin University, Geelong.

Su, S. Y. (2016). *Lara International Wetlands: How do You Conserve Migratory Bird Patterns Through the Creation of Bird Sanctuaries and Modification of Landscapes?* Unpublished SRR716 MLArch Thesis, Deakin University, Geelong.

Swan, D. (2017). *Bunya Tukka Tracks: Investigating Traditional Travelling Routes of Eastern Australia.* Unpublished MArch Thesis, Deakin University, Geelong.

Tang, B. S. S. (2013). *Cultural Project: A Design Proposal for Wathaurong Cultural Centre.* Unpublished SRL733 Assignment, Deakin University, Geelong.

Tang, B. S. S. (2014). *The Metamorphosis of Point Henry, Jillong.* Unpublished SRD768 Landscape Masterclass Project, Deakin University, Geelong.

Threadgold, H. (2020). *'What the Stones Tell Us': Gulijan Living Spaces and Landscapes.* Unpublished PhD thesis, Deakin University, Geelong.

Thurstan, R. H., Brittain, Z., Jones, D. S., Cameron, E., Dearnaley, J., & Bellgrove, A. (2018). Aboriginal Uses of Seaweeds in Temperate Australia: An Archival Assessment. *Journal of Applied Phycology, 30*(3), 1821–1832.

Wadawurrung (Wathaurung Aboriginal Corporation). (2019). *Wadawurrung Country of the Victorian Volcanic Plains*. Ballarat: Wadawurrung (Wathaurung Aboriginal Corporation).

Wagh, S. V. (2017). *Walk Amongst Us: Painkalac Creek Wetland Exploratory Trail*. Unpublished SRD768 Landscape Masterclass Project, Deakin University, Geelong.

Wagh, S. V. (2018). *Fire and Water: Drawing Initiatives from Aboriginal Theories of Fire Ecology to Manage Painkalac Creek's Wetland System at Airey's Inlet*. Unpublished SRR711 MLArch Thesis, Deakin University, Geelong.

Watson, J. (2019). *Lo-TEK: Design by Radical Indigenism*. Cologne: Taschen.

Whelan, B. (2014). *The Werribee Cultural Centre*. Unpublished SRD766 Architecture Masterclass Project, Deakin University, Geelong.

Yapa Appuhamillage, O. L. (2019). *Revival: Drug Rehabilitation Landscape – Tyers, Gippsland*. Unpublished SRD768 Landscape Masterclass Project, Deakin University, Geelong.

Yawuru Registered Native Title Body Corporate (YRNTBC). (2011). *Yawurru Cultural Management Plan: Walyjala-jala buru jayida jarringgun buru Nyamba Yawuru ngan-ga mirli mirli*. Broome: UDLA.

## CHAPTER 7

Respecting *Country* and People: Pathways Forward

David S. Jones, Kate Alder, Shivani Bhatnagar,
Christine Cooke, Jennifer Dearnaley, Marcelo Diaz,
Hitomi Iida, Anjali Madhavan Nair, Shay-lish McMahon,
Mandy Nicholson, Gavin Pocock, Uncle Bryon Powell,
Gareth Powell, Sayali G. Rahurkar, Susan Ryan,
Nitika Sharma, Yang Su, Saurabh V. Wagh,
and Oshadi L. Yapa Appuhamillage

D. S. Jones (✉) • U. B. Powell
Wadawurrung Traditional Owners Aboriginal Corporation,
Geelong, VIC, Australia
e-mail: davidsjones2020@gmail.com

K. Alder
Maribyrnong City Council, Melbourne, VIC, Australia

S. Bhatnagar • S. V. Wagh
Moir Landscape Architecture, Islington, NSW, Australia

C. Cooke • S. Ryan
School of Architecture & Built Environment, Deakin University,
Geelong, VIC, Australia
e-mail: ccooke@deakin.edu.au; rsusa@deakin.edu.au

Our Aboriginal and Torres Strait Islander tribes were the first sovereign Nations of the Australian continent and its adjacent islands, and possessed it under our own laws and customs. This our ancestors did, according to the reckoning of our culture, from the Creation, according to the common law from 'time immemorial', and according to science more than 60,000 years ago.

J. Dearnaley
Balyang Consulting, Geelong, VIC, Australia

M. Diaz
MINT Pool and Landscape Design, Melbourne, VIC, Australia
e-mail: marcediaz9940@gmail.com

H. Iida
Kihara Landscapes, Melbourne, VIC, Australia

A. M. Nair
Ground Ink, Sydney, NSW, Australia

S.-l. McMahon
GHD Woodhead, Melbourne, VIC, Australia
e-mail: Shay.McMahon@ghd.com

M. Nicholson
Tharangalk Art, Melbourne, VIC, Australia

G. Pocock
Garden Consultants, Geelong, VIC, Australia

G. Powell
Wadawurrung Traditional Owners Aboriginal Corporation,
Geelong, VIC, Australia

Legals Lawyers and Barristers (LLB) Pty Ltd, Canberra, ACT, Australia

S. G. Rahurkar
Inspiring Place, Hobart, TAS, Australia
e-mail: sayalirahurkar@gmail.com

N. Sharma
Mexted Rimmer Landscape Architecture, Geelong, VIC, Australia

Y. Su
Landscape Architect, Melbourne, VIC, Australia

O. L. Yapa Appuhamillage
Thomson Hay Landscape Architecture, Ballarat East, VIC, Australia

This sovereignty is a spiritual notion: the ancestral tie between the land, or 'mother nature', and the Aboriginal and Torres Strait Islander peoples who were born therefrom, remain attached thereto, and must one day return thither to be united with our ancestors. This link is the basis of the ownership of the soil, or better, of sovereignty. It has never been ceded or extinguished, and co-exists with the sovereignty of the Crown. (extract from the *Uluru Statement from the Heart*, Referendum Council 2017)

Key ethos of contemporary Aboriginal culture in Australia includes a desire for sovereignty recognition, a request for equality of voice in decision-making as it pertains to their respective *Country* and people with the capacity for people to 'Ask first', that theirs is a culture that co-lives in the past, present and future and that all are infinite, and a desire to share part or whole of their Indigenous Knowledge Systems subject to respect and 'asking'. Also, a recognition that they are the custodians of more than 60,000 years of environmental knowledge pertinent to this landscape, its lands, its waters, its skies, and its portent moods.

While many built environment academics and practitioners are 'lost' in approaching this topic realm applicable to their discipline, citing apprehensiveness, lack of resources and expertise, lack of internal and external Aboriginal expertise to guide and 'teach' (Jones et al. 2017; Tucker et al. 2016), this book demonstrates that possibilities and exemplars exist (Jones et al. 2018). It is the younger generations that are increasingly thirsty for guidance, engagement protocols, and to appreciate where they respectively 'sit' inside this continent, or inside their home *Country*.

To move forward, as a built environment professional practitioner, in this situation is not just to be scaffolded with Western technical and theoretical knowledges and skills, but to additionally appreciate its longitudinal picture and its innate nuances and personality. To appreciate and talk with the latter is respect. It includes navigating their ethical obligations to heal this continent and its waters arising from the last 200 years of colonisation but also to understand what needs to be additionally considered in land use, environmental management, design and planning decision-making and creating. It is not simply that we need to 'train' the current generation of architects, landscape architects and planners in Western expectations, but that we need to nurture a respectful and healing ethic in them—irrespective of whether it is the Australian landscape, the Indian landscape, the Chinese landscape, the Indonesian landscape, and so on—that enhances their global citizenship credentials. The latter is what universities and Australian built environment institutes are ironically desirous of ensuring that are embodied in the values of their graduates and practitioners respectively.

REFERENCES

Jones, D. S., Low Choy, D. G, Revell, G., Heyes, S., Tucker, R., & Bird, S. (2017). *Re-casting Terra Nullius Blindness: Empowering Indigenous Protocols and Knowledge in Australian University Built Environment Education.* Canberra, ACT: Office for Learning and Teaching / Commonwealth Department of Education and Training. At http://www.olt.gov.au/project-re-casting-terra-nullius-blindness-empowering-indigenous-protocols-and-knowledge-australian- (2017–2018); at https://ltr.edu.au/resources/ID12_2418_Jones_Report_2016.pdf (2018), (accessed 1 May 2020).

Jones, D. S., Low Choy, D., Tucker, R., Heyes, S., Revell, G., & Bird, S. (2018). *Indigenous Knowledge in the Built Environment: A Guide for Tertiary Educators.* Canberra, ACT: Office for Learning and Teaching / Commonwealth Department of Education and Training; https://ltr.edu.au/resources/ID12-2418_Deakin_Jones_2018_Guide.pdf (accessed 1 May 2020).

Tucker, R., Low Choy, D., Heyes, S., Revell, G., & Jones, D. S., (2016). Re-Casting terra nullius design-blindness: Better teaching of Indigenous Knowledge and protocols in Australian architecture education, *International Journal of Technology and Design Education, 28*(1), 303–322.

Referendum Council. (2017). *Uluru Statement from the Heart.* Canberra: Referendum Council. Available at: https://ulurustatement.org/our-story. Accessed 1 May 2020.

INDEX

A

Aboriginal and Torres Strait Islander peoples, 3
Aboriginal Architecture, 2
Aboriginal concepts of *Country*, 12
Aboriginal Heritage Act 2006, 66
Aceh, 73
Acknowledgement of Country, 51
Adnyamathanha, 67
AIA Victoria Reconciliation Action Plan, 50
Ainu, 73
Allwood, B., 98
Andrews, T. D., 30
Arabana, 65
Arbon, V., 29, 65
Archaeoastronomy, 67
Architects Accreditation Council of Australia's (AACA), 48
Architects Board of Western Australia, 48
Architects Institute of Australia (AIA), 46
Architects Registration Board of Victoria, 48
Architectural Practice Board of South Australia, 48
Architecture, 2
Arrente, 67
Australian Capital Territory Architects Board, 48
Australian High Court, 23
Australian Institute of Landscape Architects (AILA), 6, 46
Australian Local Government Association (ALGA), 21
Australian Qualifications Framework, 63
Autoethnological, 6
Awabakal, 66

B

Bali Aga, 73
Barry, J., 29
Basso, K. H., 67
Battiste, M., 29
Bell, Damein, 66
Benterrak, K., 67
Bhatnagar, S., 98

Bielawski, E., 30
Bishop, R., 29
Boandik, 23
Board of Architects of Queensland, 48
Board of Architects of Tasmania, 48
Bodkin-Andrews, G., 22
Brabham, Wendy, 65
Brambuk Cultural Centre, 67
Breeden, S., 67
Briggs, Carolyn, 66
Brown, P., 102
Brunner, J., 51
Buggey, S., 13
Burarrwanga, Laklak, 16
Bush tukka, 67
Byrne, J., 51

C
Cahir, F., 102
Calendars, 67
Canada, 25
Canadian Indigenous context, 13
Carbyn, L., 30
Casault, A., 33
Chambers, C. D., 49
Chilisa, B., 29
China, 71
City of Greater Geelong Council, 66
Clarke, P. A., 66
Climate change, 7
Closing the Gap, 21
Closing the Gap Reports, 21
CODA Studio, 24
Colonisation, 22
Colorado, P., 30
Connection to Country Policy, 51
Constitution (Scheduled Castes) Order, 1950, 69
Constitution (Scheduled Tribes) Order, 1950, 71
Cooke, C., 51, 102

Country, v, 7
Crosby, A., 32
Cultural landscapes, 13
Cultural sustainability, 33

D
Dalla Costa, W., 25
Davidson, J., 33
Dawson, J., 67
Deakin, Alfred, 61
Deakin University (Deakin), 6, 61
Deakin University Act 1974, 61
Deakin University Act 2009, 62
Deakin University Human Research Ethics Committee (DUHREC), xiii
Declaration on the Rights of Indigenous Peoples, 13
Decolonialism, 3
Decolonised design, 3
Decolonised education experiences, 3
Deconstruction, 26
Degar, 73
Designing narratives, 98
Dhudhuroa, 65
Djillong, 65
Dovey, K., 67
Dreaming, 16
Dyirrbal gumbilbara bama, 24

E
Eccles, C., 95
8 Ways of Learning, 27
Ellis, C., 29
Eora, 66
Erickson, M., 50
Ethics, 26
Ethnobotany, 67
eVALUate, 80

F

Fantin, S., 33, 34
Far North Queensland Regional Plan 2009–2031, 32
Fien, J., 33
Findley, L., 67
First Nation, 25
First Nations peoples, 7
Flannery, T., 67
Flood, J., 67
Food and Agriculture Organisation, 30
Fortin, David, 25
Fourmile, G. G., 33, 34
Freeman, M., 30
Frontier Wars, 3

G

Gagudju, 67
Gammage, B., 23
Gardiner, D., 32
Gariwerd/The Grampians, 67
Geelong, 65
Gerritsen, R., 67
Gimuy, 34
Glasson, J., 51
Go-Sam, C., 24
Gott, B., 67
Graduate Learning Outcomes (GLOs), 32, 63
Grant, E., 33
Guidelines for Ethical Research in Australian Indigenous Studies, 66
Guiding Principles for the Development of Indigenous Cultural Competency in Australian Universities, 31–32
Gunai/Kurnai, 66
Gunditjmara, 65
Gunditjmara People with Wettenhall, 102
Gurran, N., 51

H

Heyes, S., 50
Hmong (Miao), 73
Hokkaidō, 73
Houston, S., 49
Hromek, D., 16
Hughes, P., 29
Hull-Milera, V., 29

I

India, 67, 69
Indigenous Architecture and. Design Victoria (IADV), 50
Indigenous Content in Education Symposium, 32
Indigenous cultural competency, 32
Indigenous Knowledge Systems (IKS), vi, 2, 6, 8, 22
Indigenous peoples' perceptions of spaces, 16
Indigenous sovereignty, 23
Indigenous view of environment, 7
Indigenous world view, 7
Intercultural design, 33
International Council for Science (ICSU), 30
International Labour Organisation, 30

J

Jackson, S., 23
Japan, 67
Jones, D. S., 3, 30, 33, 95, 98

K

Keys, C., 33
Kickett-Tucker, C., 24
Kiddle, R., 33
Kovach, M., 28
Krog, S. R., 33

L

Landscape architecture, 2
Language, 26
Lawson, G., 50
Liddle, C., 30
Lochert, M., 48
Lourandos, H., 67
Low Choy, D., 33

M

Mabo Case, 23
*Mabo v the State of Queensland
    [No. 2]*, 30
Maginn, P. J., 51
Mak Mak people, 15
Malaysia, 67
Mallie, T., 48
Malnar, J. M., 32
Martin, K., 27
Martin, T., 33
Massola, A., 67
McBryde, I., 67
McGaw, J., 32
Meintangk, 23
Memmott, P., 32, 33
Memmott, P. C.,
    48, 49
Merculieff, I., 30
Métis Nation, 25
Minahasan, 73
Minorities Rights Group
    International, 30
Mirraboopa, M., 27
Mitchell, N., 13
Moran, C., 29
Moran, M., 33
More, A. J., 29
Morieson, J., 67
Murphy, M., 50
Murujuga Aboriginal Corporation
    (MAC), 102

N

Nader, L., 30
National Architecture Conference, 49
National Assessment Program–Literacy
    and Numeracy (NAPLAN), 21
National Coalition of Aboriginal and
    Torres Strait Islander Peak
    Organisations (NCATSIPO), 21
National Indigenous Australians
    Agency (NIAA), 21
*National Standard of Competency for
    Architects*, 48
Native Title Act 1993 (Cth), 23
Neale, M., 102
Neidjie, B., 67
New Zealand, 29
Nganampa Health Council Inc., 49
Ngarrindjeri, 66
Nicholson, M., 98
Noonuccal, 27
Northern Territory Architects
    Board, 48
NSW Architects Registration
    Board, 48
Nyeri Nyeri, 65

O

Oberklaid, S., 51
O'Brien, Lewis, 66
Olukanni, D., 32
Orang Asli, 69
Ostwald, M., 48
OutReach, 50

P

Palawa, 16
Paradies, Y., 65
Parks Victoria, 102
*The Partnership Agreement on Closing
    the Gap, 2019–2029*, 21

Pascoe, B., 23
Phibbs, P., 51
Pholeros, P., 49
Pieris, A., 32
Place-making, 7, 98
Plangermaireener, 34
Planning Institute of Australia (PIA),
    8, 46, 51
Pocock, G., 95
Porsanger, J., 29
Porter, L., 23, 29
Powell, B., 24
Professional accreditation, 3

Q
Quandamookah, 34

R
Rahurkar, S. G., 98
Ramanujan, A. K., 29
Recasting Terra Nullius Blindness, 3
Reconciliation, 3
Reconciliation Action Plans, 32
Rees, S. L., 16
Reser, J., 48
Revell, G., 25
Rigney, L. –I., 29
Roös, P. B., 95
Rose, D. B., 15
Rose, Mark, 29, 30, 65
Ryan, S., 98

S
Saddle Lake Cree, 25
Sanyal, S., 29
Sarkissian, W., 51
Sauer, C., 13
Sawyer, A., 48
Scally, S., 49

Scheduled Castes (SCs), 69
Scheduled Tribes (STs), 69
Seasonal calendars, 73
Seasons, 67
Self-determination, 26
Sen, Geeti, 29
Sharma, N., 102
Sheehan, N. W., 29
Sherwood, J., 24
Sinatra, J., 50
Small, G., 32
Smith, L. T., 26
Spurr, S., 50
SRL733 Indigenous Narratives and
    Processes (SRL733), xiii, 6
Stallard, K. E., 49
Steffensen, V., 102
Stewart, P., 32
Stilgoe, J., 13
Student Evaluation of Teaching and
    Units (SETU), 80
Su, R. J., 98
Sulawesi, 73
Sumatra, 73
Sustainable design, 32
Swan, D., 98
Sweet, M. A., 23

T
Tang, B. S. S., 98
Tanganekald, 23
Tasmania, 34
Tawa, M., 67
Tay, 73
Ted, G., 29
Terrestrial and aquatic
    ethnobotany, 95
Tertiary Education Quality and
    Standards Agency (TEQSA), 47
Tharoor, S., 29
Threadgold, H., 98

Thurstan, R. H., 98
Tobias, T., 29
Tonkinson, R., 49
Tournier, David, 66
Tucker, R., 3, 33
Tuiteci, S., 50
Tunbridge, D., 67

U
Uluru, 67
Uluru-Kata Tjuta Cultural Centre, 67
Uluru Statement from the Heart, 115
United Nations Educational, Scientific
    and Cultural Organisation
    (UNESCO), 13
Unit Learning Outcomes
    (ULOs), 7, 67
Universities Australia, 30
University of Western Australia, 50
Urban planning, 2

V
Venice Architecture Biennale, 25
Victoria, 51
Vietnam, 73
Vodvarka, F., 32

W
Wadawurrung (Wathaurung
    Aboriginal Corporation)
    (W(WAC)), 66
Wadawurrung Country, 7, 66, 95
Wadjuk, 24
Wagh, S. V., 98

Waldram, J., 30
Walker, S., 29
Walliss, J., 33
Wamba Wamba, 65
Ward, M., 49
Wathaurong Aboriginal Co-Operative
    (WAC), 66
Wathaurung Aboriginal
    Corporation, xiii
Watson, I., 23
Ways of Knowing, Being and Doing, 27
Wells, K., 32
Wensing, E., 32
Wergaia, 65
Whelan, B., 98
Williams, P., 49
Wilson, S., 29
Wiradjuri, 24
Wong, K., 24
World Conference on Science, 30
World Health Organisation, 30
Wurundjeri, 66

Y
Yapa Appuhamillage, 98
Yawurru, 67
Yawuru Registered Native Title Body
    Corporate (YRNTBC), 102
Yorta Yorta, 66
Yuin, 16
Yunkaporta, T., 27

Z
Zao, 73
Zola, N., 67

Printed in the United States
by Baker & Taylor Publisher Services